PRAC
PROBL
MATHEMATICS
for MANUFACTURING

Delmar's *PRACTICAL PROBLEMS in MATHEMATICS* Series

- *Practical Problems in Mathematics for Automotive Technicians, 4e*
  George Moore
  Order # 0-8273-4622-0

- *Practical Problems in Mathematics for Carpenters, 5e*
  Harry C. Huth
  Order # 0-8273-4579-8

- *Practical Problems in Mathematics for Electricians, 5e*
  Herman and Garrard
  Order 0-8273-6708-2

- *Practical Problems in Mathematics for Electronic Technicians, 5e*
  Herman and Sullivan
  Order # 0-8273-6761-9

- *Practical Problems in Mathematics for Graphic Artists*
  Vermeersch and Southwick
  Order # 0-8273-2100-7

- *Practical Problems in Mathematics for Heating and Cooling Technicians, 2e*
  Russell B. DeVore
  Order # 0-8273-4062-1

- *Practical Problems in Mathematics for Manufacturing, 4e*
  Dennis D. Davis
  Order # 0-8273-6710-4

- *Practical Problems in Mathematics for Masons, 2e*
  John E. Ball
  Order # 0-8273-1283-0

- *Practical Problems in Mathematics for Mechanical Drafting*
  John C. Larkin
  Order # 0-8273-1670-4

- *Practical Problems in Mathematics for Welders, 4e*
  Schell and Matlock
  Order # 0-8273-6706-6

Related Titles

- *Mathematics for the Automotive Trade, 2e*
  Peterson and DeKryger
  Order # 0-8273-3554-7

- *Mathematcs for Electricity and Electronics*
  Dr. Arthur Kramer
  Order # 0-8273-5804-0

PRACTICAL PROBLEMS in MATHEMATICS for MANUFACTURING

4th edition

Dennis D. Davis

Delmar Publishers

ITP An International Thomson Publishing Company

Albany • Bonn • Boston • Cincinnati • Detroit • London • Madrid • Melbourne
Mexico City • New York • Pacific Grove • Paris • San Francisco • Singapore • Tokyo
Toronto • Washington

NOTICE TO THE READER

Publisher does not warrant or guarantee any of the products described herein or perform any independent analysis in connection with any of the product information contained herein. Publisher does not assume, and expressly disclaims, any obligation to obtain and include information other than that provided to it by the manufacturer.

The reader is expressly warned to consider and adopt all safety precautions that might be indicated by the activities herein and to avoid all potential hazards. By following the instructions contained herein, the reader willingly assumes all risks in connection with such instructions.

The publisher makes no representation or warranties of any kind, including but not limited to, the warranties of fitness for particular purpose or merchantability, nor are any such representations implied with respect to the material set forth herein, and the publisher takes no responsibility with respect to such material. The publisher shall not be liable for any special, consequential, or exemplary damages resulting, in whole or in part, from the readers' use of, or reliance upon, this material.

Cover Credit: Dartmouth Publishing

Delmar Staff:
Publisher: Michael A. McDermott
Editor: Mary Clyne
Production Manager: Larry Main
Art & Design Coordinator: Nicole Reamer

Printed in the United States of America

For more information, contact:
Delmar Publishers
3 Columbia Circle, Box 15015
Albany, New York 12212-5015

International Thomson Publishing Europe
Berkshire House 168 - 173
High Holborn
London WC1V 7AA
England

Thomas Nelson Australia
102 Dodds Street
South Melbourne, 3205
Victoria, Australia

Nelson Canada
1120 Birchmount Road
Scarborough, Ontario
Canada M1K 5G4

International Thomson Editores
Campos Eliseos 385, Piso 7
Col Polanco
11560 Mexico D F Mexico

International Thomson Publishing GmbH
Königswinterer Strasse 418
53227 Bonn
Germany

International Thomson Publishing Asia
221 Henderson Road
#05 - 10 Henderson Building
Sinapore 0315

International Thomson Publishing - Japan
Hirakawacho Kyowa Building, 3F
2-2-1 Hirakawacho
Chiyoda-ku, Tokyo 102
Japan

3 4 5 6 7 8 9 10 xxx 01 00 99 98 97

Library of Congress Cataloging-in-Publication Data

Davis, Dennis D.
Practical problems in mathematics for manufacturing/
Dennis D. Davis -- 4th ed.
p. cm.-- (Delmar's practical problems in mathematics series)
ISBN: 0-8273-6710-4
1. Shop mathematics. 2. Industrial engineering--Mathematics
I. Title II. Series
TJ1165.D26 1995
513'. 14' 02467--dc20

95-3236
CIP

Contents

APPENDIX

Preface

Many products are designed to be constructed of steel, aluminum, or other materials such as plastics. If the products themselves are not constructed of these materials, the tools and machines used to manufacture them are surely made from metals and plastics. The machine operators in the production and tool shops build these machines as well as the products themselves. *Practical Problems in Mathematics for Manufacturing* provides the student with realistic mathematical problems that manufacturers commonly encounter. By solving the problems, both technical and mathematical skills are strengthened, providing a solid foundation for a career in manufacturing.

Practical Problems in Mathematics for Manufacturing is one of a series of books that can be used in conjunction with a comprehensive mathematics textbook or in an individualized mathematics program. Each workbook provides practical exercises in using mathematical principles to solve occupationally related problems. This edition provides Critical Thinking Problems at the end of each section. Most of these problems will require the student to draw on lessons previously learned and may require the student to do some basic research to solve the problem. The majority of the Critical Thinking Problems have been drawn from the experience of the author and are representative of problems technicians and engineers must deal with on a daily basis in industry.

This series of workbooks is designed for use by a wide range of students. They are suitable for any student from high school to the two-year college level. The series has many benefits for the instructor and for the student. For the student, the workbooks offer a step-by-step approach to mastery of essential skills in mathematics. Each workbook includes relevant and easily understandable problems in a specific vocational field. For the instructor, the series offers a coherent and concise approach to the teaching of mathematical skills. Each workbook is complemented by an Instructor's Guide, which includes answers to every problem in the workbook and solutions to many of the problems. In addition, two achievement reviews are provided at the end of the Instructor's Guide to provide an effective means of measuring students' progress. Both the instructor and the student will benefit from the specific vocational material and the appendix materials, such as performing the basic operations with denominate numbers.

Dennis Davis has more than twenty years of experience in the manufacturing industry serving as a tool and die maker, automated machine designer, technical college instructor, manufacturing engineer, and plant manager. He is certified as a manufacturing engineering technologist by the Society of Manufacturing Engineers. Mr. Davis is currently Industrial Engineering Manager for Armstrong Rim & Wheel Manufacturing in Armstrong, Iowa.

Whole Numbers

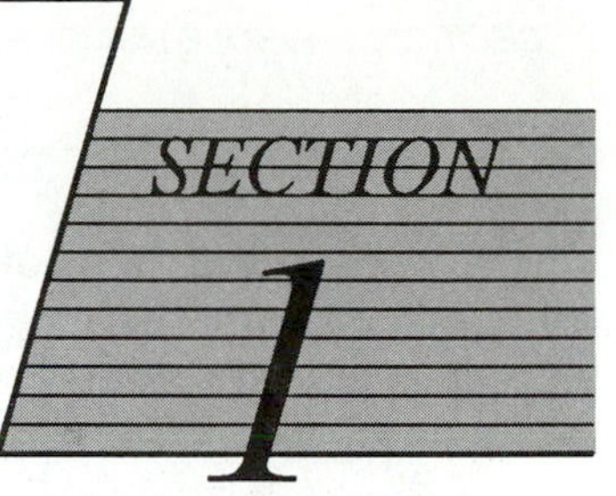

Unit 1 ADDITION OF WHOLE NUMBERS

BASIC PRINCIPLES AND PROCEDURES

Addition is the process of adding numbers to create a sum. Numbers are placed one over the other with columns aligned vertically by units, tens, hundreds, and so on.

123	The units column is added first and the "carry" is added to column
4567	2. Then column 2 is added and the "carry" is added to column 3,
+ 89	and so on.

3 + 7 + 9 = 19 Bring the 9 down to become the units digit and carry the one to column 2.

1 + 2 + 6 + 8 = 17 Bring the 7 down to become the tens digit in the answer and carry the 1 to column 3.

1 + 1 + 5 = 7 Bring the seven down to become the hundreds digit of the answer. Bring the 4 down to become the thousands digit of the answer. The answer is 4779.

REVIEW PROBLEMS

Add the following quantities:

a.	b.	c.	d.	e.
13 mm	400 sq. in.	1,659 ft.	$1,200	47,111 km
33 mm	890 sq. in.	13,482 ft.	999	3,134 km
125 mm	87 sq. in.	672 ft.	5,872	256 km
109 mm	233 sq. in.	93 ft.	56	9,989 km

PRACTICAL PROBLEMS (Note: No allowance is made for the saw kerf.)

1. Calculate the mass of the six machines shown. ________

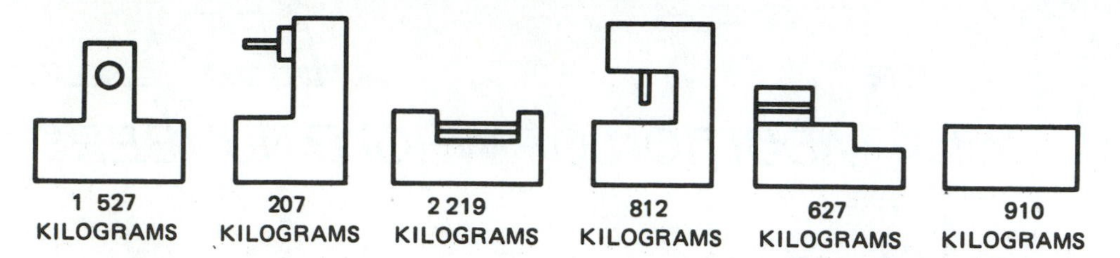

2. The following lengths of high-speed steel were cut off a six-meter bar: 984 mm (millimeters), 763 mm, 127 mm, 812 mm, 904 mm, 429 mm, 99 mm, 84 mm, and 637 mm. Determine the total length, in millimeters, that has been cut off. ________

3. Determine the total number of bars required if a screw machine uses the following materials: 42 bars of 25 mm stock, 249 bars of 20 mm stock, 377 bars of 15 mm stock, 928 bars of 12 mm stock, and 1,015 bars of 8 mm stock. ________

4. Calculate the number of cutters on the following stock room cutter rack: 5 side milling cutters, 17 slitting saws, 23 end milling cutters, and 49 angle cutters. ________

5. Find distances *A, B, C,* and *D.*

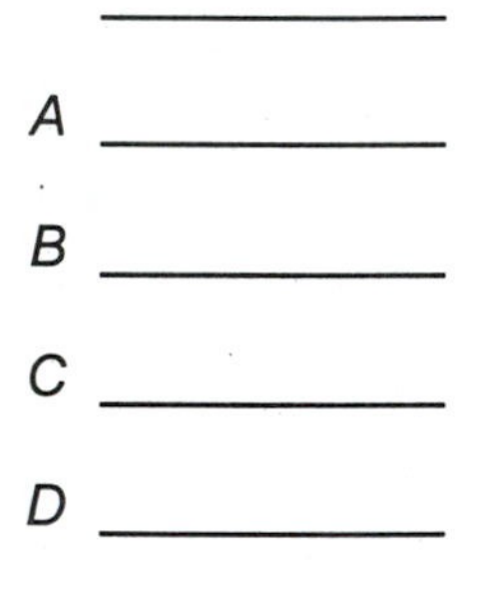

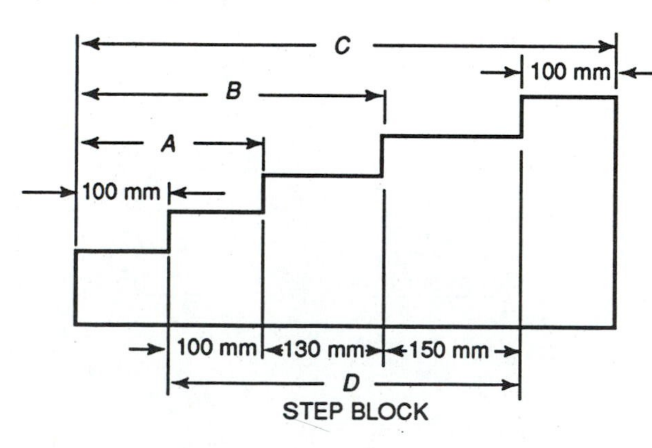

STEP BLOCK

6. Find the length of *A*, *B*, *C*, and *D*.

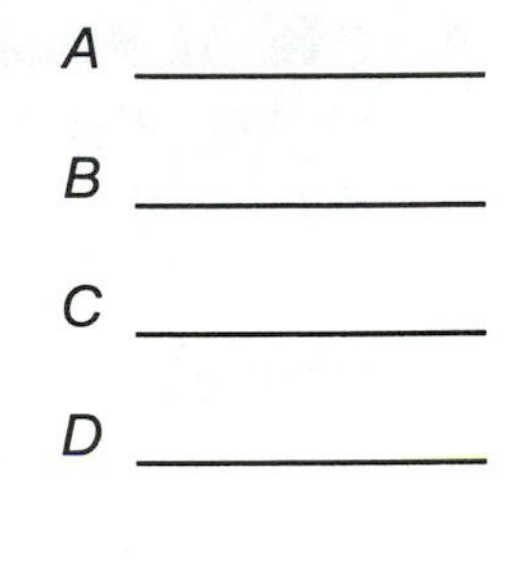

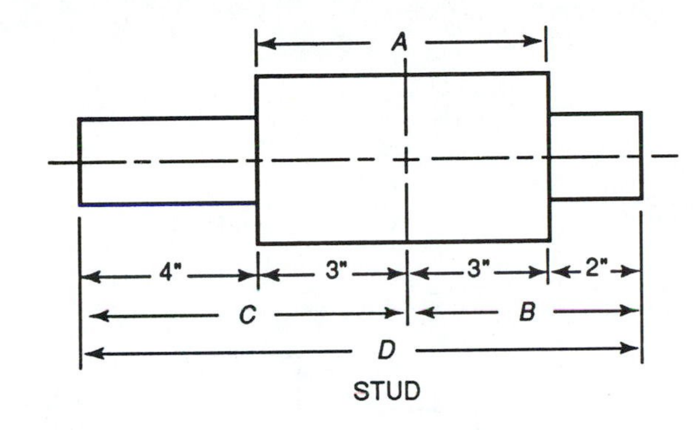

STUD

7. Five 10-mm (millimeter) holes are drilled in a 10 mm by 40 mm by 150 mm piece of flat stock.

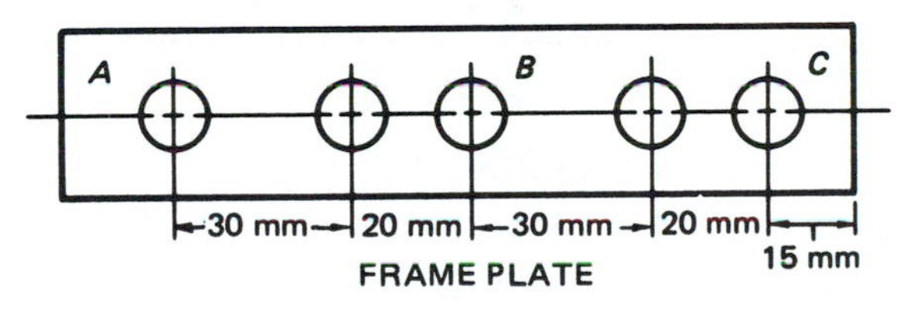

FRAME PLATE

a. Determine the total length between centers of holes *A* and *C*. a. ____________

b. Determine how far hole *B* is from the right end. b. ____________

c. Calculate the distance hole *A* is from the right end. c. ____________

8. Provide answers to the following questions concerning the metric bolt shown.

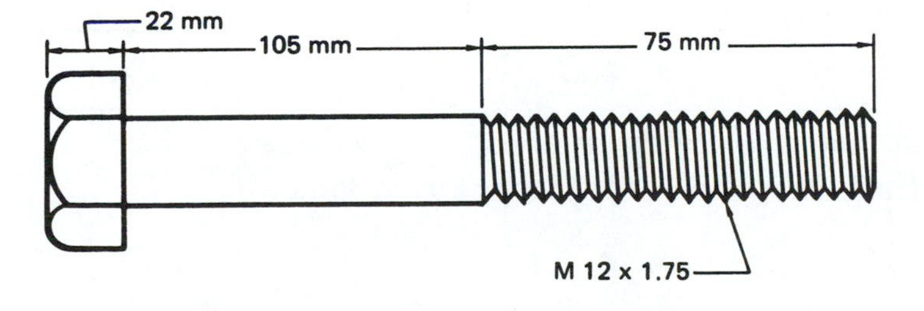

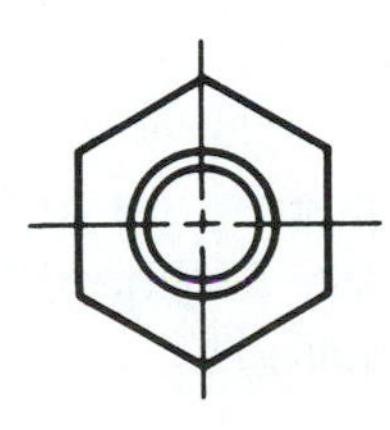

a. Find the overall length. a. ____________

b. What is the length of the threaded portion? b. ____________

c. What is the thickness of the head? c. ____________

d. Find the length of the bolt excluding the head. d. ____________

9. Five numbered shafts, like the one shown, vary in size. The chart lists the dimensions of each shaft in inches.

SHAFT	A	B	C	D	E	X
1	1	3	7	5	2	
2	2	4	15	12	2	
3	5	14	9	13	2	
4	1	3	16	12	1	
5	12	2	13	3	2	

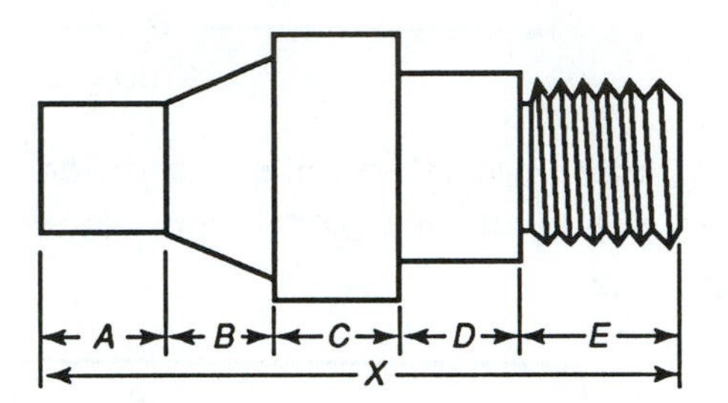

a. Using the measurements given in the chart determine the overall length *X* of each shaft in inches.

a. shaft *1* __________

shaft *2* __________

shaft *3* __________

shaft *4* __________

shaft *5* __________

b. What is the total length of the five shafts?

b. __________

10. Several pieces are to be cut from a 1-inch diameter rod. The lengths needed are 7 inches, 14 inches, 31 inches, and 3 inches. How long a rod is required?

____

Unit 2 SUBTRACTION OF WHOLE NUMBERS

BASIC PRINCIPLES AND PROCEDURES

Subtraction is the process of finding the numerical difference between numbers.

$$\begin{array}{r} 1236 \\ -\ \ 54 \\ \hline \end{array}$$

The problem is set up as in addition with the columns aligned vertically. Start by subtracting 4 from 6. 6 - 4 = 2 The 2 becomes the units digit in the answer. Since 3 is smaller than 5 we cannot do this subtraction without "borrowing" a ten from the 100's column. When we borrow the ten from the 100's column, the 2 in the 100's column becomes a 1. The 3 in the 10's column becomes a 13. Now we can subtract 5 from the 13. 13 - 5 = 8. The 8 becomes the 10's digit of the answer. Since there are no more numbers in the bottom number to subtract, simply bring the ones in the 100's column and the 1000's column down to become part of the answer. The answer is 1182.

REVIEW PROBLEMS

Subtract the following quantities:

a.	b.	c.	d.	e.
156 cm	1,564 yd.	13,621 oz.	$48,363	123,452 ft.
- 34 cm	- 198 yd.	- 5,329 oz.	- 19,482	-120,986 ft.

PRACTICAL PROBLEMS (Note: No allowance is made for the saw kerf.)

1. Determine how much stock remains on a rack containing 476 meters of 30-mm stock if a worker uses 289 meters. ____________

2. Determine how many millimeters of stock remain on a 2,000-mm bar if the following lengths are cut off the bar: 892 mm, 122 mm, 22 mm, and 107 mm. 30 mm are lost to saw kerf. ____________

3. Calculate oil remaining in the tank if a worker removes the amounts shown. ____________

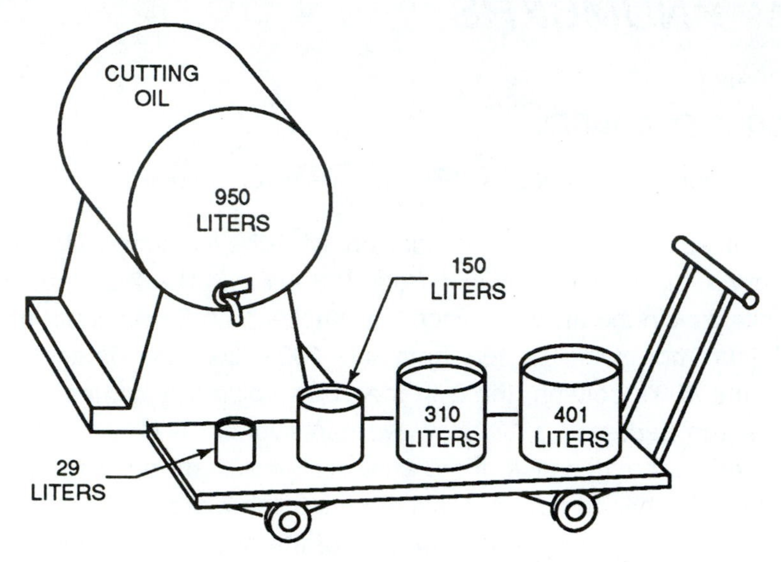

4. What is the size of the diameter of the hole in the washer? ____________

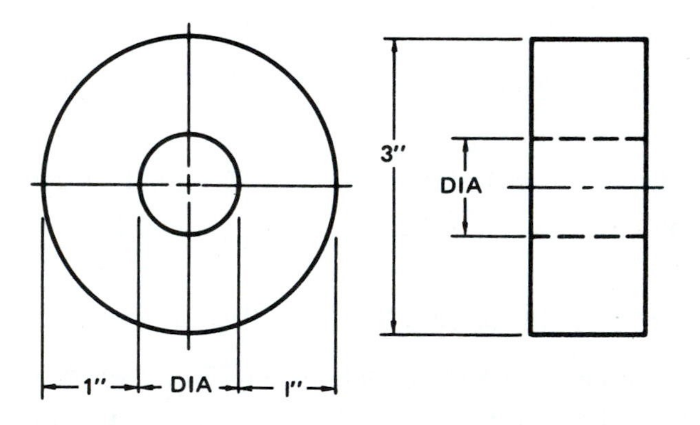

Note: Use this diagram for problems 5 and 6.

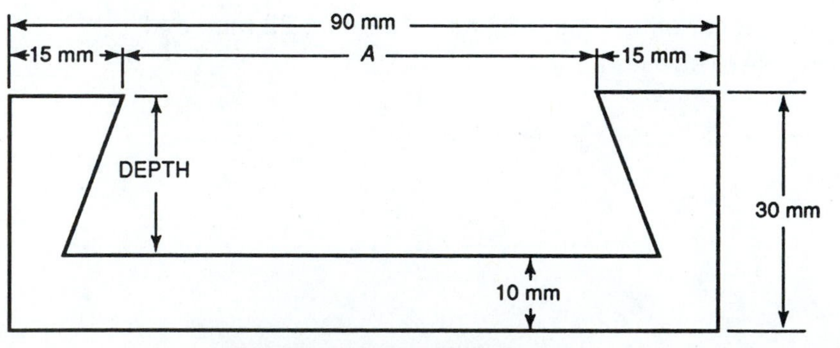

FEMALE DOVETAIL BLOCK

5. Find distance A. ________

6. Determine the depth of the dovetail. ________

Note: Use this diagram for problems 7 and 8.

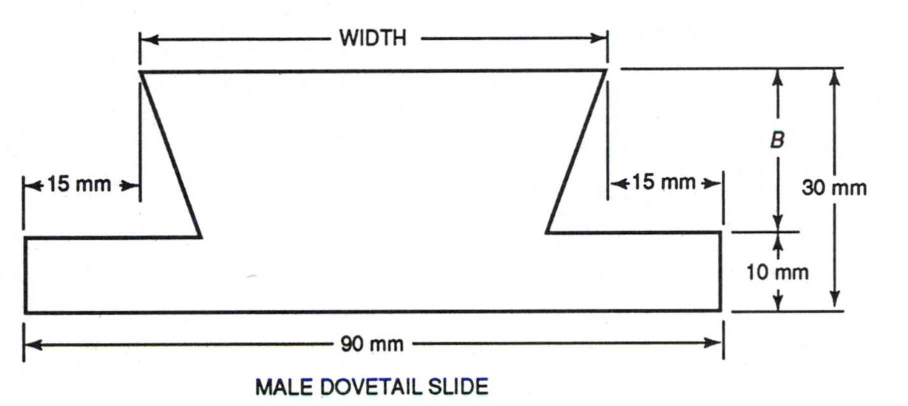

MALE DOVETAIL SLIDE

7. Find the width across the top of the slide. ________

8. Find dimension B. ________

9. Find the lengths of *A, B, C,* and *D.*

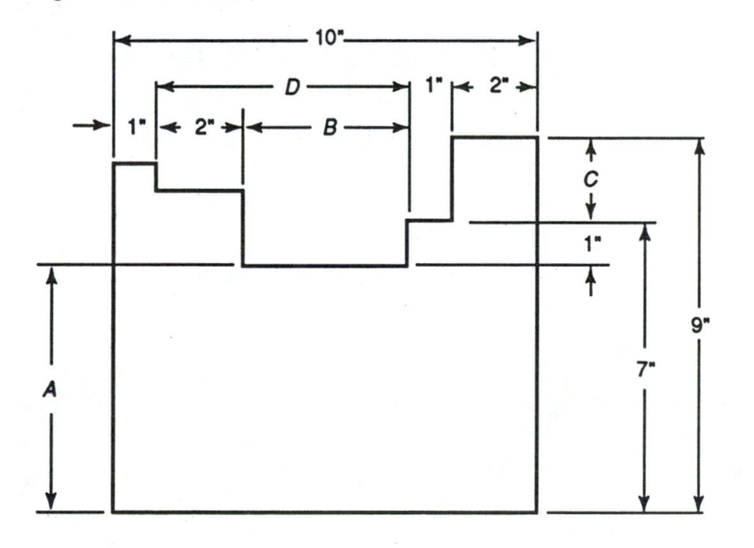

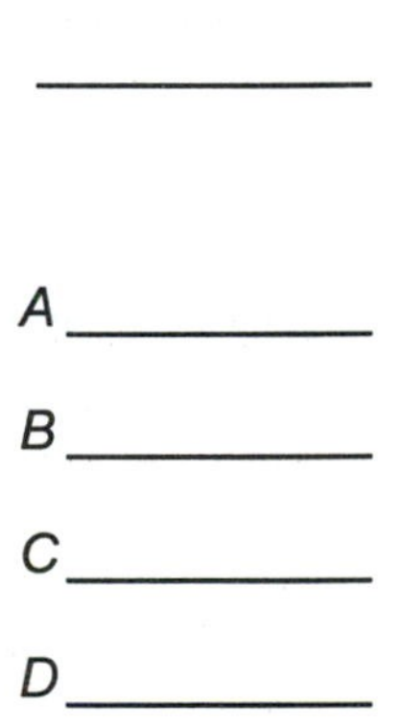

10. A gear on a splined shaft is shown. The dimensions of four similar parts are shown in millimeters. Determine the width ***X*** of the gear on each of the four parts.

GEAR	*A*	*B*	*C*	*D*	*X*
1	20	60	30	160	
2	30	120	20	200	

Gear *1* ____________

Gear *2* ____________

Gear *3* ____________

Gear *4* ____________

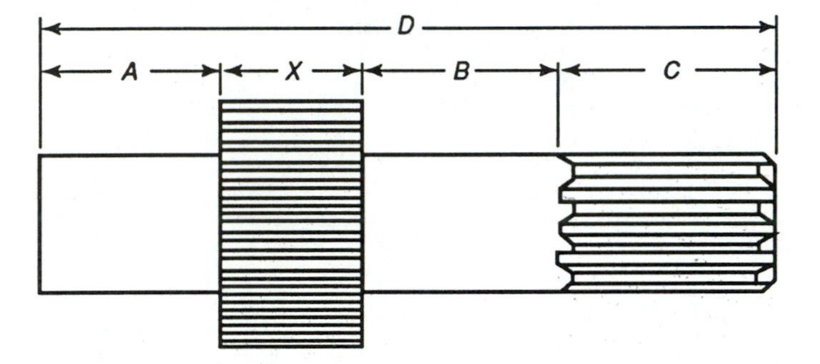

11. Five shafts with a total length of 109 inches are cut from one 144-inch length of stock. How much is left? ____________

Unit 3 MULTIPLICATION OF WHOLE NUMBERS

BASIC PRINCIPLES AND PROCEDURES

Multiplication is essentially successive addition. For example, if we do the following addition problem:

$$6 + 6 + 6 + 6 = 24$$

It is the same as doing the multiplication problem:

$$6 \times 4 = 24$$

Multiplication is much easier and less error prone. Doing multiplication requires that multiplication tables be memorized or a calculator be used.

Example Problem:

$$\begin{array}{r} 29 \\ \underline{\times\ 12} \\ 58 \end{array}$$

First multiply 29 x 2. Carries are used in multiplication as in addition. Place the first partial answer, 58, below the line.

Now multiply 29 x 1. Remember that we are actually multiplying by 10, so when we bring the second partial answer down, we move it one place to the left.

$$\begin{array}{r} 29 \\ \underline{\times\ 12} \\ 58 \\ \underline{29} \\ 348 \end{array}$$

Add the two partial answers to obtain the final answer.

Note that this problem is the same as adding 29 to itself 12 times. Multiplication makes solving the problem easier, faster, and less error prone.

REVIEW PROBLEMS

Multiply the following quantities:

a. $\begin{array}{r} 93 \\ \underline{\times\ 12} \end{array}$ b. $\begin{array}{r} 986 \\ \underline{\times\ 44} \end{array}$ c. $\begin{array}{r} 2{,}896 \\ \underline{\times\ 309} \end{array}$ d. $\begin{array}{r} \$987 \\ \underline{\times\ \ 40} \end{array}$ e. $\begin{array}{r} \$18{,}350 \\ \underline{\times\ \ \ \ 88} \end{array}$

PRACTICAL PROBLEMS (Note: No allowance is made for the saw kerf.)

1. Determine how much steel rod is required to make 23 bolts like the one shown. A total of 5 inches is used for cutoff and chucking. __________

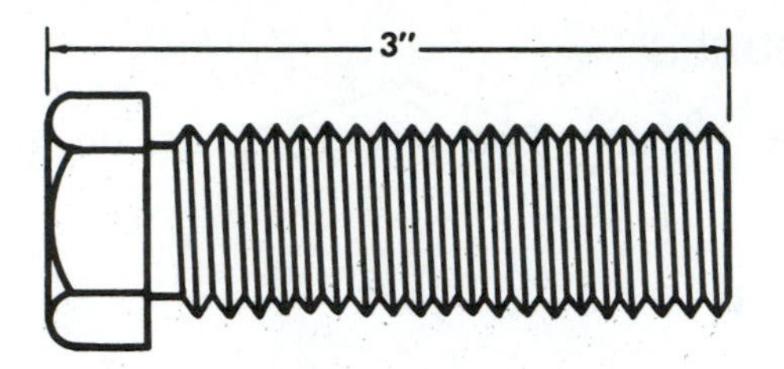

2. What length of bar is needed to cut 17 pins, each 19 millimeters long? (Allow a total waste of 50 millimeters for cutting.) __________

3. Find the total length of each size of material that follows:

a.	7 each	25 mm diameter x 6 m (meters) long	steel rod	a. __________
b.	17 each	30 mm diameter x 9 m long	aluminum rod	b. __________
c.	8 each	22 mm diameter x 3 m long	aluminum rod	c. __________
d.	29 each	10 mm diameter x 11 m long	brass rod	d. __________

4. Find the number of threads there are on a threaded rod 7 inches long with eight threads per inch (TPI). __________

5. A job requires 3 hours lathe work, 2 hours drilling, 5 hours surface grinding and 7 hours on a milling machine. If 235 similar jobs are ordered, determine the total number of hours that would be required for each of the processes.

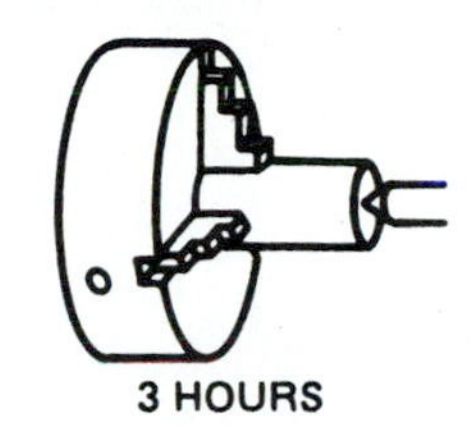

3 HOURS

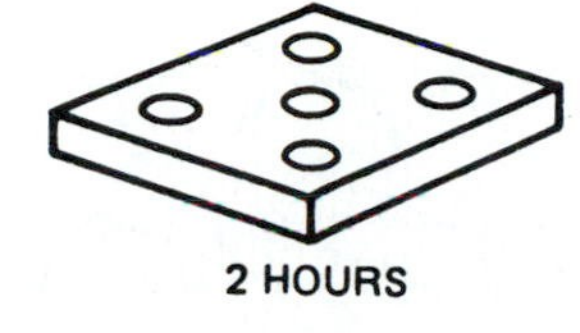

2 HOURS

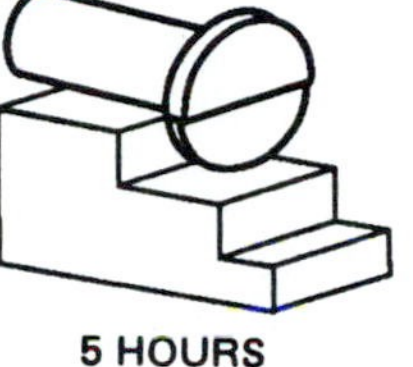

5 HOURS

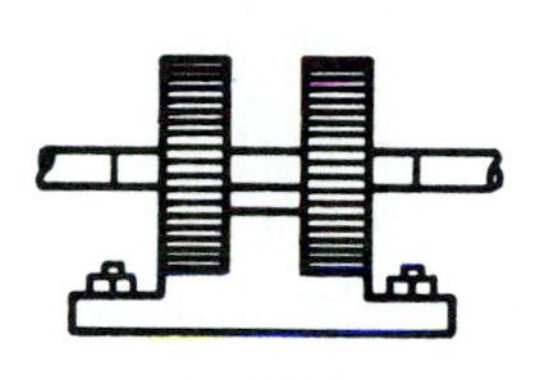

7 HOURS

   a. Lathe — a. ____________

   b. Drill press — b. ____________

   c. Surface grinder — c. ____________

   d. Milling machine — d. ____________

6. A machinist uses 456 bundles of 15-millimeter bar stock. Each bundle contains 27 bars. What is the total number of bars used? ____________

7. Each of 16 milling machines weighs 2,478 pounds. What is the total weight of these machines? ____________

8. Subtract 4,735 from 7,623 and then multiply the difference by 658. What is the product? ____________

9. A plant produces 2,324 spindles on each shift. If the plant operates two shifts a day, five days a week, how many spindles are produced in four weeks? ____________

10. Determine the total length of rod required to make nine of the shafts shown. ______

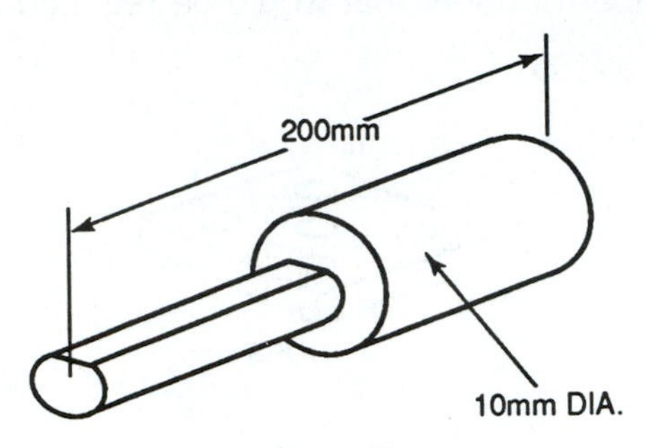

11. On a certain lathe the stock turns 16 times for the tool to advance 1 inch. How many times must the stock turn for the tool to advance 3 inches? ______

Unit 4 DIVISION OF WHOLE NUMBERS

BASIC PRINCIPLES AND PROCEDURES

Division is essentially successive subtraction. If we were to subtract 6 from 36, we would find that we could perform that operation 6 times before we had zero remaining. Again, division requires the memorization of multiplication tables because multiplication is used in division. A division problem may be checked easily by using multiplication. For example:

$$250 \div 5 = 50 \qquad \text{Check: } 50 \times 5 = 250$$

Example Problem:

```
     39
 9 ) 356
     27
     86
     81
   r = 5
```

Try to divide 9 into 3. Since 3 is too small to be divided evenly by 9, try dividing 9 into 35. 9 will go into 35 3 times. Place 3 over the 5 and then multiply 3 x 9, which is 27. Place this below the 35.

Subtract 27 from 35 to obtain 8, and bring the 8 down. Since 8 cannot be divided by 9, bring down the 6, and divide 86 by 9.

$86 \div 9 = 9$ with 5 left over. The final answer to this problem is then 39 r 5 or $39\frac{5}{9}$.

REVIEW PROBLEMS

Divide the following quantities:

a. $30 \div 5 =$ ________ b. $4{,}044 \div 22 =$ ________ c. $550 \div 34 =$ ________

d. $625 \div 5 =$ ________ e. $8{,}845 \div 789 =$ ________ f. $999 \div 33 =$ ________

PRACTICAL PROBLEMS

1. A machinist has a piece of 256-millimeter long stock which is cut into eight equal pieces. How long is each piece? ________

2. A machine shop owner spends $2,368 to buy tools. Each tool costs an average of $16. How many tools can be bought? ________

3. If 252,636 pieces are made in 12 hours, determine the number of pieces that are machined per hour. ________

4. An automatic screw machine uses 96 millimeters for each cycle of operation. How many pieces can be made from 240 bars of stock, each 8 meters (8,000 millimeters) long? ________

5. To produce a part, an automatic screw machine uses 8 inches of stock, including the cutoff. How many complete parts can be machined from a 248-foot bar? HINT: 1 foot = 12 inches ________

6. A punch press stamps out 4,200 small discs per hour. How many hours does it take to produce 252,000 parts? ________

7. It takes a turret lathe operator 12 minutes to produce a part. How many parts are produced in 8 hours? HINT: 1 hour = 60 minutes. ________

8. Four holes are drilled in the plate shown. The center-to-center distance between holes is equal to the distance from the center of each end hole to the end of the plate. Find center-to-center distance X between the holes. ________

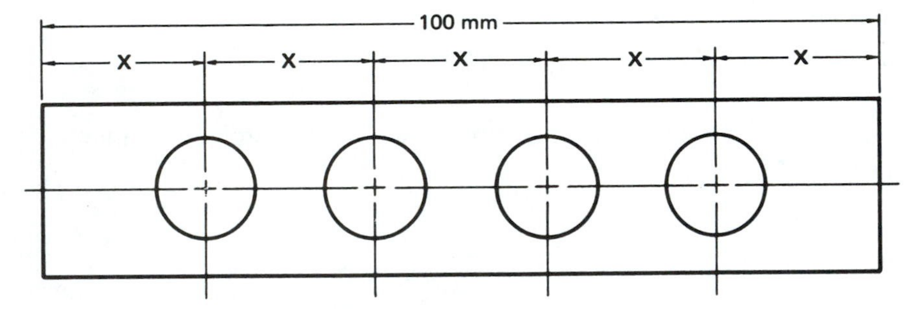

9. How many pins like the one shown can be cut from a 2-meter bar of stock? HINT: 1 meter = 1,000 millimeters ________

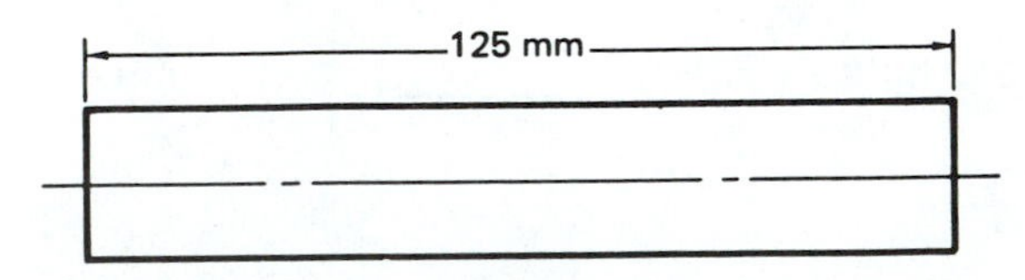

10. How many reamers can be made from a piece of stock 4 feet long? ________

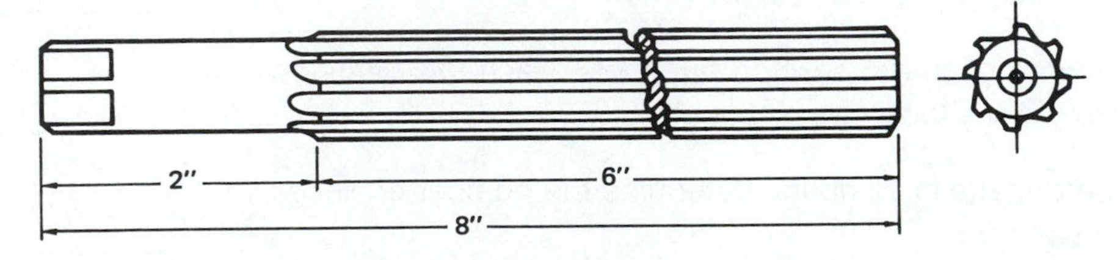

11. How many 200-millimeter-long reamers can be made from a 3-meter-long bar of tool steel? HINT: 1 meter = 1,000 millimeters __________

12. How many reamers, each 250 millimeters long, can be made from a 4-meter-long bar of tool steel? HINT: 1 meter = 1,000 millimeters __________

13. A spring weighs 250 grams. How many coiled springs can be made from a 20-kilogram bundle of wire? HINT: 1 kilogram = 1,000 grams. __________

14. From a bar of steel 19 feet long, 72 bolts are made. One foot of stock is lost in cutting off and machining. What is the length of each bolt? __________

15. Four dozen twist drills cost $192.

    a. What is the cost per dozen? a. __________

    b. What does each drill cost? b. __________

CRITICAL THINKING PROBLEMS

1. A tool room supervisor must send five parts out to be heat treated. Each part weighs 6.3 pounds and the shipping company charges $5.00 for the first 10 lbs. of the shipment plus $0.35 for each pound over 10 lbs. Determine the shipping cost for the five parts. (Round answer to the nearest cent.) __________

2. Determine the following dimensions:

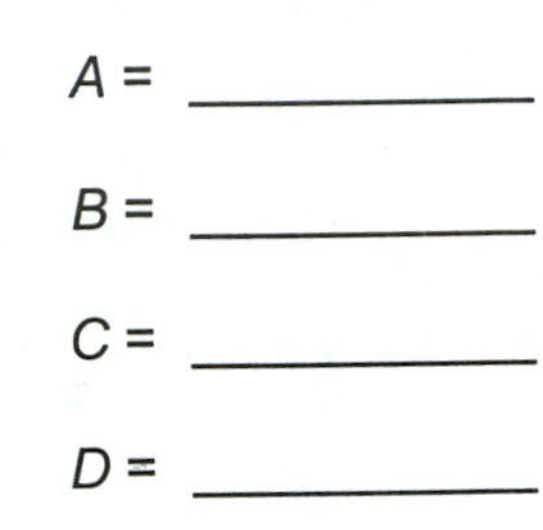

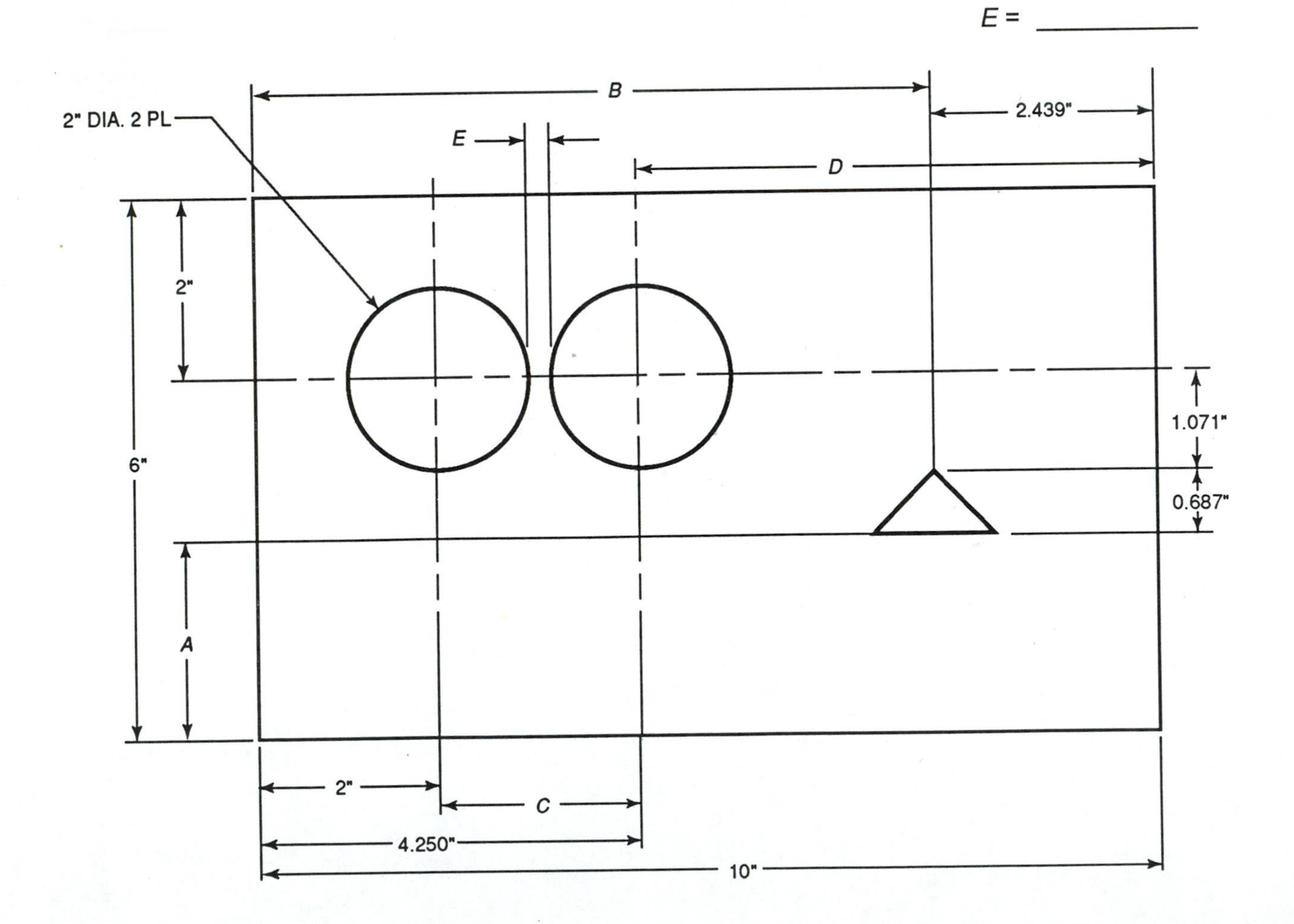

Common Fractions

Unit 5 ADDITION OF COMMON FRACTIONS

BASIC PRINCIPLES

- *Fractions* are used to divide units into equal parts. For example, an inch could be evenly divided into 32 equal parts and each part would be $\frac{1}{32}$ of an inch. In the machine trade it is common to use fractions as small as $\frac{1}{64}$ of an inch. There are two parts of a fraction. The *numerator* is the number above the line and the *denominator* is the number below other line. To add fractions that have the same denominator, simply add the numerators together to form the numerator of the answer. The denominator of the answer will simply be the common denominator of the fractions added.

- It is a common practice to reduce fractions to lowest terms by dividing both the numerator and the denominator by the same number until both are reduced to the point that neither can be reduced any further. Note that in the following example both the four and the eight can be divided by four, leaving one half.

$$\frac{3}{8} + \frac{1}{8} = \frac{4}{8} = \frac{1}{2} \text{ ans.}$$

- When fractions to be added do not have the same denominators, one must be determined by finding the lowest number into which each denominator can be divided evenly. This number is referred to as the *lowest common denominator.*

$$\frac{1}{2} = \frac{8}{16} \qquad \text{LCD} = 16$$

$$+ \quad \frac{3}{8} = \frac{6}{16}$$

$$\frac{14}{16} = \frac{7}{8} \text{ ans.}$$

- Sometimes a fraction is formed where the numerator is larger than the denominator. In this case the fraction can be written as a mixed number by dividing the numerator by the denominator. For example, the fraction $^{12}/_{5}$ can be written as $2^{2}/_{5}$. The number of times the denominator goes into the numerator evenly will become the whole number portion of the mixed number. If there is a remainder, it becomes the numerator of the fractional part of the mixed number.

REVIEW PROBLEMS

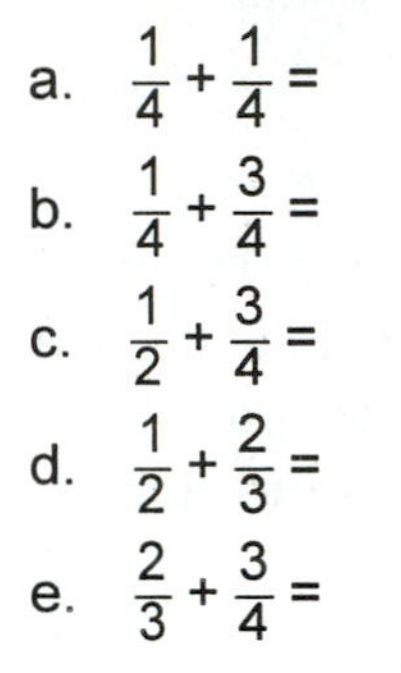

a. $\frac{1}{4} + \frac{1}{4} =$

b. $\frac{1}{4} + \frac{3}{4} =$

c. $\frac{1}{2} + \frac{3}{4} =$

d. $\frac{1}{2} + \frac{2}{3} =$

e. $\frac{2}{3} + \frac{3}{4} =$

PRACTICAL PROBLEMS

Note: Use this diagram and information for problems 1–5.
When making a machine part some material is wasted in operations such as facing and cutting the steel from the bar. In making this shaft, $^{1}/_{8}$ inch is allowed for waste.

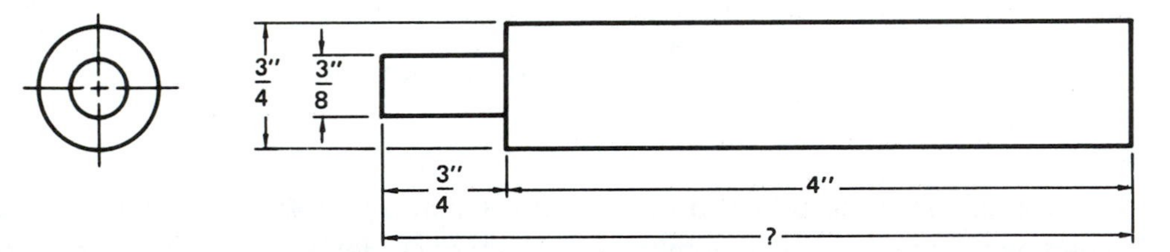

1. What length of stock is required to make this shaft? __________

2. How long is the finished shaft? __________

3. If the small end of the shaft is changed to $1^{1}/_{8}$ inches long and all other dimensions remain the same, what is the overall length? __________

4. When the length dimensions are $5^{1}/_{2}$ inches and $^{15}/_{16}$ inch, what is the overall length? __________

5. How long a piece of steel is needed to make a shaft with length dimensions of $5^{1}/_{2}$ inches and $^{15}/_{16}$ inch? __________

6. The length under the head of a bolt is the length from the underside of the head to the threaded end. In this bolt, it is the length of the turned and threaded part or the length of the bolt without the head. Find the length under the head of the bolt. __________

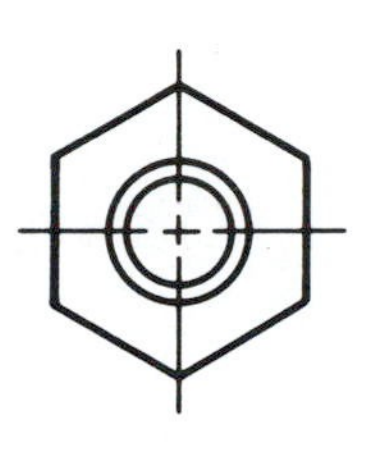

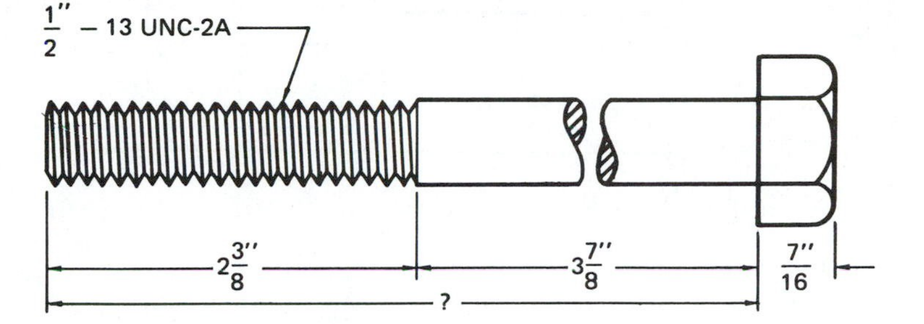

Note: Use this diagram for problems 7 and 8.

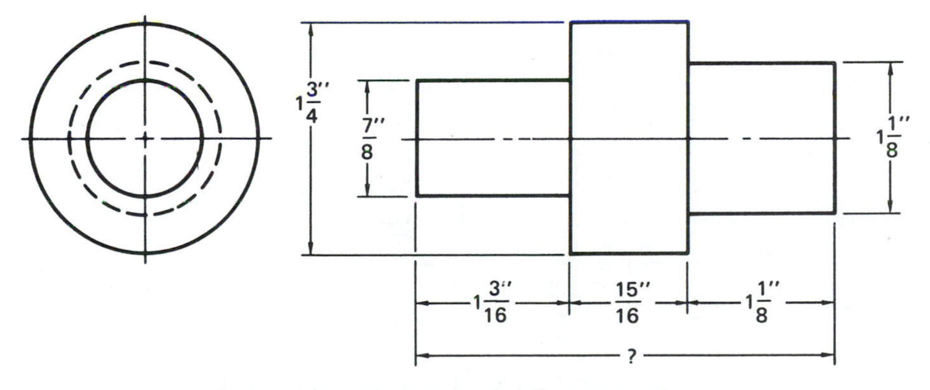

7. What is the overall length of the arbor as illustrated? __________

8. A similar arbor has dimensions of $^{17}\!/_{32}$ inch, $^{13}\!/_{16}$ inch, and $^{21}\!/_{32}$ inch. Allowing $^{1}\!/_{8}$ inch for waste, find the length of stock needed to make this arbor. __________

9. Five pieces of steel are cut from a bar. The lengths are $^{7}\!/_{8}$ inch, $^{9}\!/_{16}$ inch, $^{3}\!/_{4}$ inch, $^{27}\!/_{32}$ inch, and $^{15}\!/_{16}$ inch.

   a. What is the total length of the five pieces? a. __________

   b. What is the total length of stock needed to cut the five pieces? Allow $^{1}\!/_{8}$ inch for waste on each of the pieces. b. __________

10. Find the overall length of this shaft. ____________

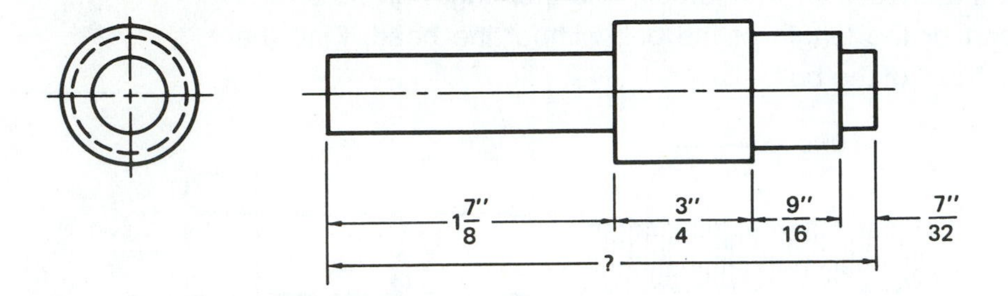

11. What length of stock will be used in making this link? Allow $\frac{1}{8}$ inch for the cutoff saw and for finishing the ends. ____________

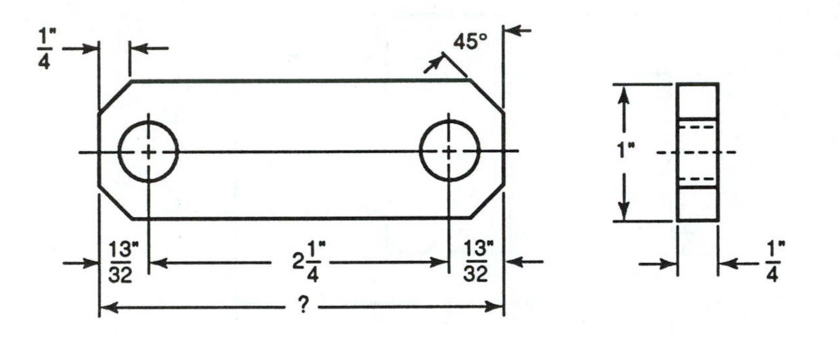

12. This sketch shows a profile gauge that is used for checking work in a machine shop. The dimensions on the sketch are the lengths of the different steps. Find the overall length of the gauge. ____________

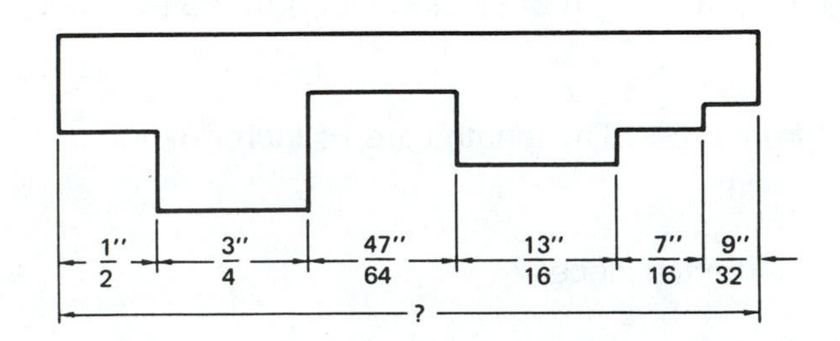

Note: Use this diagram for problems 13–15.

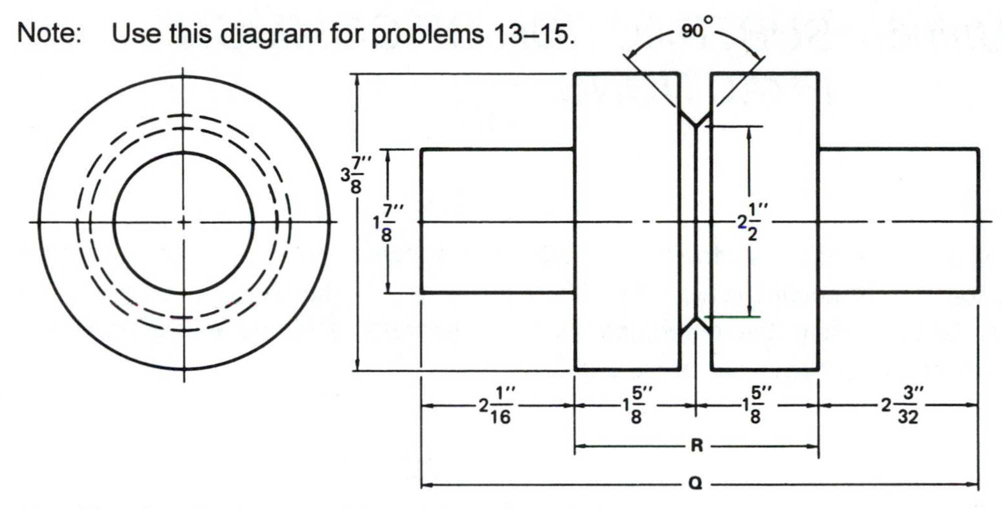

13. How long is the gear blank (dimension ***Q***)?

14. How long is the center portion (dimension ***R***)?

15. The dimensions of length for the gear blank are changed to $1\frac{3}{8}$ inches, $2\frac{11}{32}$ inches, $2\frac{11}{32}$ inches, and $1\frac{57}{64}$ inches. Allowing $\frac{1}{8}$ inch for waste, what is the length of stock needed for this job?

Note: Use this diagram for problems 16–18.

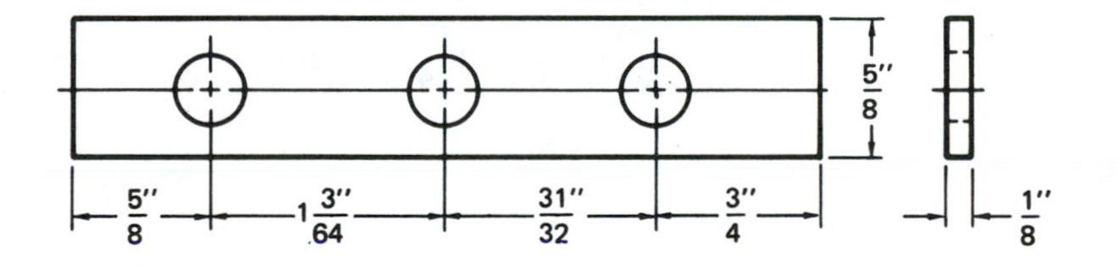

16. What is the center-to-center distance between the outside holes?

17. What is the overall length of the plate?

18. If $\frac{1}{32}$ inch is allowed on each end for finishing, what is the length of stock used?

Unit 6 SUBTRACTION OF COMMON FRACTIONS

BASIC PRINCIPLES

- As in addition of fractions, subtraction requires that a *lowest common denominator* be found. Once this is accomplished, the difference in the numerators is found and a fraction is formed using the difference as the numerator and the *lowest common denominator* as the denominator of the answer.

$$\frac{9}{32} = \frac{18}{64} \qquad \text{LCD} = 64$$

$$-\frac{3}{64} = \frac{3}{64}$$

$$\frac{15}{64} \text{ ans.}$$

The rules for mixed numbers apply for subtraction also.

$$5\frac{3}{8} = \frac{43}{8} = \frac{129}{24} \qquad \text{LCD} = 24$$

$$-1\frac{2}{3} = \frac{5}{3} = \frac{40}{24}$$

$$\frac{89}{24} = 3\frac{17}{24} \text{ ans.}$$

REVIEW PROBLEMS

a. $\frac{1}{2} - \frac{1}{4} =$

b. $\frac{1}{2} - \frac{1}{8} =$

c. $\frac{3}{4} - \frac{2}{3} =$

d. $\frac{1}{8} - \frac{3}{32} =$

e. $\frac{7}{8} - \frac{7}{32} =$

PRACTICAL PROBLEMS

Note: Use this diagram for problems 1–3.

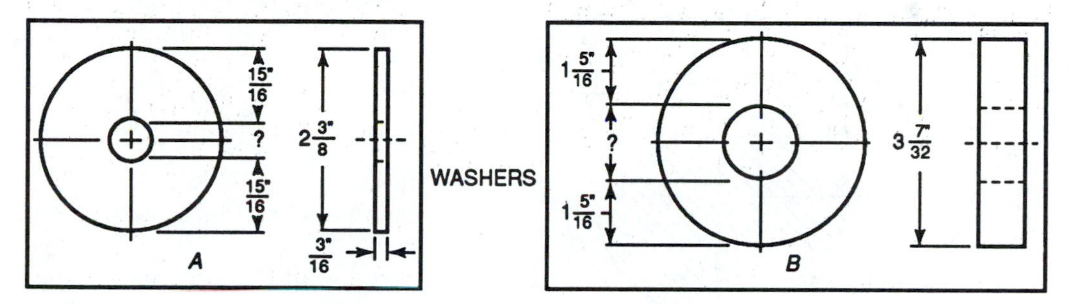

1. A machinist receives a work request for these two washers. What is the inside diameter of washer *A*? __________

2. The outside diameter of washer *A* remains the same. The wall thickness is changed from $^{15}/_{16}$ inch to $^{13}/_{32}$ inch. What diameter hole will this washer have? __________

3. What is the hole diameter in washer *B*? __________

Note: Use this diagram for problems 4 and 5.

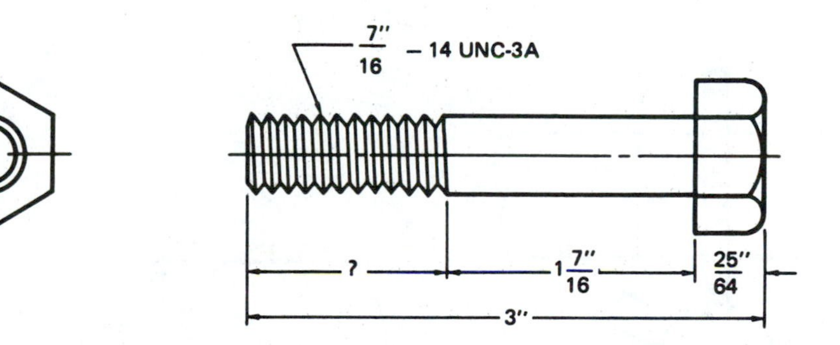

4. Ten bolts, like the one shown, must be made for an assembly. Find the length under the head on the bolt. __________

5. Find the length of the threaded part on this bolt. __________

Note: Use this diagram for problems 6 and 7.

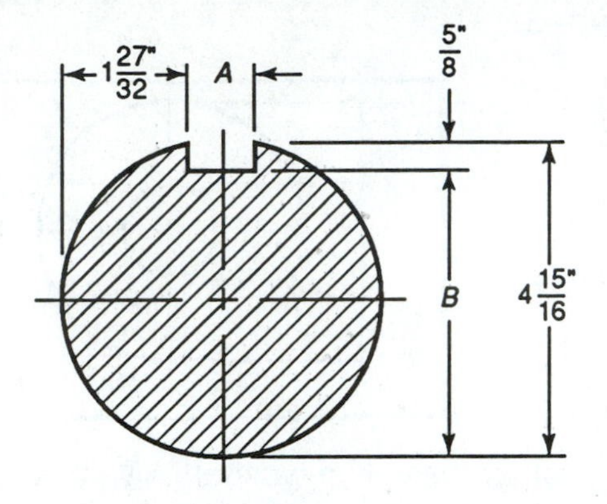

6. A machinist must mill this keyway. To do this properly, certain dimensions must be known. Find the width of the keyway in the motor shaft (dimension *A*). __________

7. What is the distance from the bottom of the keyway to the opposite side of the shaft (dimension *B*)? __________

8. A keyway in a $2\frac{1}{4}$-inch shaft is milled to a depth of $\frac{5}{16}$ inch. What is the measurement from the bottom of the keyway to the opposite side of the shaft? __________

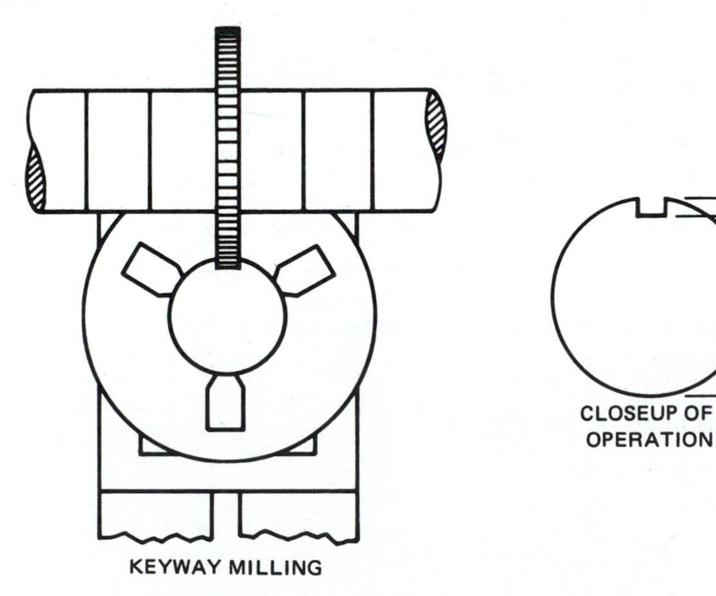

9. The measurement from the side of a $2\frac{1}{2}$-inch shaft to the edge of the keyway is $1\frac{1}{64}$ inches. Find the width of the keyway. ______

Note: Use this diagram for problem 10.

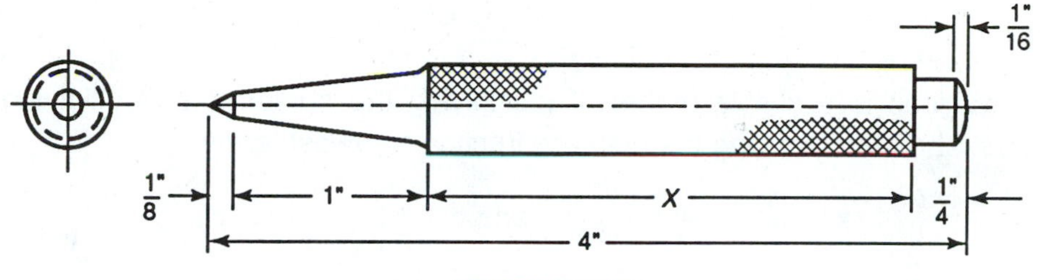

CENTER PUNCH

10. The illustrated center punch is made from a piece of tool steel. Before making the punch, the toolmaker must know certain dimensions. What is the dimension X on the illustrated center punch? ______

Note: Use this diagram for problems 11 and 12.

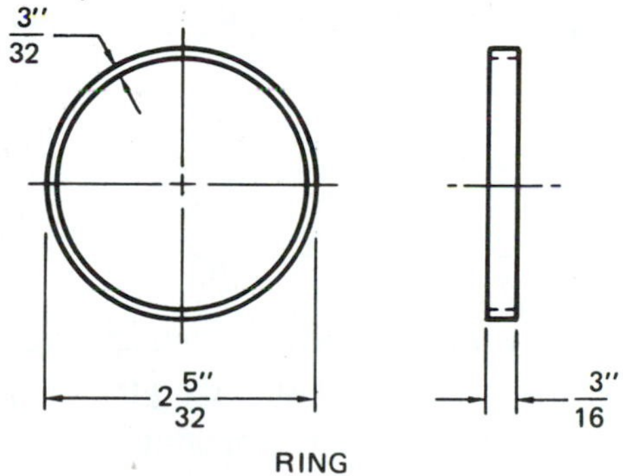

RING

11. A machinist receives a work request and drawing for the illustrated ring. Find the inside diameter of the ring. ______

12. If the wall thickness is $\frac{1}{16}$ inch and the outside diameter is $1\frac{19}{64}$ inches, find the inside diameter of the ring. ______

13. If the outside diameter of a pipe measures $2\frac{1}{8}$ inches and the wall thickness is $\frac{9}{32}$ inch, what is the inside diameter? ______

14. On a lathe, a $\frac{9}{32}$ inch cut is made from $1\frac{1}{4}$-inch diameter stock. What is the finished diameter? ______

15. A bar of cast iron, $22\frac{1}{8}$ inches long, has three pieces cut from it. The pieces measure $6\frac{1}{2}$ inches, $4\frac{7}{8}$ inches, and $2\frac{5}{32}$ inches in length. If $\frac{1}{8}$ inch is allowed for each saw cut, how long a piece of stock is left? ______

Unit 7 MULTIPLICATION OF COMMON FRACTIONS

BASIC PRINCIPLES

- Multiplying fractions is a simple matter of multiplying the numerators, multiplying the denominators, and then reducing the resulting fraction to lowest terms.

$$\frac{7}{8} \times \frac{1}{2} = \frac{7}{16} \text{ ans.}$$
$$\frac{3}{4} \times \frac{2}{3} = \frac{6}{12} = \frac{1}{2} \text{ ans.}$$

- Sometimes a fraction is formed with a numerator that is larger than the denominator. In this case, the fraction is written as a mixed number by dividing the numerator by the denominator. The number of times the denominator goes into the numerator evenly is the whole number portion of the mixed number. If there is a remainder, it becomes the numerator of the fractional part of the whole number. The original denominator is then used as the new denominator.

$$\frac{3}{2} \times \frac{12}{9} = \frac{36}{18} = 2 \text{ ans.}$$
$$\frac{7}{8} \times \frac{3}{2} = \frac{21}{16} = 1\frac{5}{16} \text{ ans.}$$

- Converting a mixed number to a fraction requires multiplying the denominator by the whole number portion of the mixed number and adding the result to the numerator. The original denominator is then used as the new denominator.

$$2\frac{1}{5} = \frac{11}{5} \text{ ans.}$$

REVIEW PROBLEMS

Multiplying the following fractions:

a. $\frac{1}{2} \times \frac{1}{4} =$

b. $\frac{1}{3} \times \frac{2}{3} =$

c. $\frac{3}{5} \times \frac{5}{16} =$

d. $\frac{5}{6} \times \frac{12}{5} =$

e. $\frac{17}{32} \times \frac{60}{2} =$

PRACTICAL PROBLEMS

1. Using $5\frac{7}{8}$ inches for each chisel, what is the shortest bar that can be used for making five chisels? __________

2. A chisel requires $4\frac{9}{16}$ inches of $\frac{1}{2}$-inch hexagonal stock. How long a bar is needed to make 25 of these chisels? __________

3. Each chisel requires $4\frac{5}{16}$ inches of stock. What length of $\frac{1}{2}$-inch hexagonal steel is used in making 17 chisels? __________

4. If each chisel weighs $1\frac{3}{4}$ pounds, what is the weight of 12 chisels? __________

5. How long a piece of $\frac{13}{16}$-inch hexagonal bar is needed to make 12 $\frac{1}{2}$-inch nuts? Allow $\frac{1}{16}$ inch for waste on each nut. __________

6. If $\frac{3}{32}$ inch is allowed for waste for each nut, how much 1-inch hexagonal stock is needed for 25 $\frac{5}{8}$-inch standard nuts? __________

7. Nuts are made from a 1 $\frac{1}{8}$-inch hexagonal bar. The hexagonal bar is 20 feet long and $\frac{3}{32}$ inch is allowed for waste for each nut. How much bar is left after making 50 $\frac{3}{4}$-inch nuts? __________

Note: Use this diagram for problems 8 and 9.

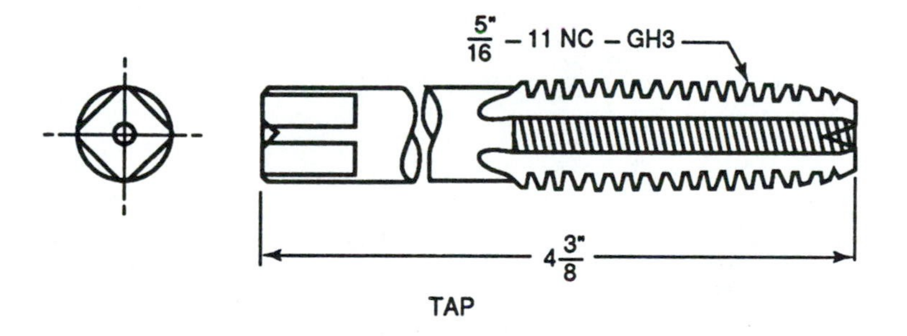

TAP

8. A toolmaker receives a work request to make the tap shown. If $\frac{1}{16}$ inch is allowed for finishing the ends of each tap, what length of stock is needed to make two taps? __________

9. What length of stock is needed for $\frac{1}{2}$ dozen taps? Allow $\frac{1}{16}$ inch for finishing both ends of each tap. __________

10. At $87\frac{1}{2}$ cents each , how much do 12 washers cost? __________

11. A hacksaw blade has 14 teeth to the inch. The toothed section is 9 inches long. A machinist uses only $4\frac{1}{4}$ inches of toothed section.

    a. How many teeth are actually doing the cutting? a. ____________

    b. How many teeth are not being used? b. ____________

Unit 8 DIVISION OF COMMON FRACTIONS

BASIC PRINCIPLES

- Dividing fractions requires inverting the fractional divisor, and multiplying numerators and denominations as in multiplication of fractions. The resulting fraction is then written as a mixed number.

$$\frac{3}{4} \div \frac{1}{2} = \quad \text{(Invert the } {}^{1}\!/_{2}\text{)}$$

$$\frac{3}{4} \times \frac{2}{1} = \frac{6}{4} = 1\frac{2}{4} = 1\frac{1}{2} \text{ ans.}$$

REVIEW PROBLEMS

a. $\frac{1}{2} \div \frac{1}{3} =$

b. $\frac{3}{4} \div \frac{4}{5} =$

c. $\frac{1}{3} \div \frac{3}{4} =$

d. $\frac{17}{32} \div \frac{3}{5} =$

e. $\frac{7}{17} \div \frac{6}{10} =$

PRACTICAL PROBLEMS

Note: Use this diagram for problem 1.

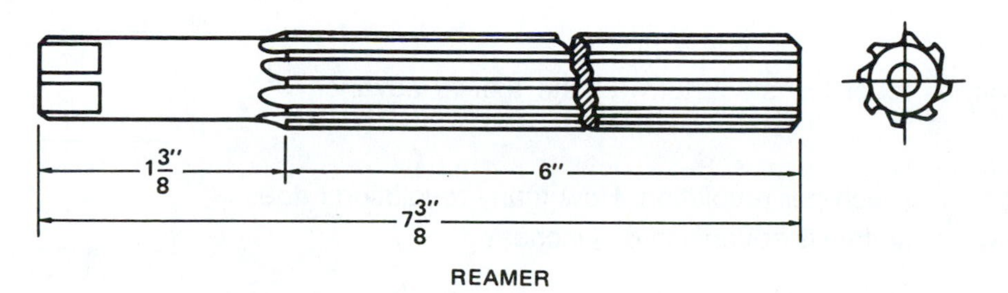

REAMER

1. How many reamers can be made from a piece of stock 3 feet long? Allow ⅛ inch on each reamer for waste. __________

Note: Use this information for problems 2 and 3. The *pitch* of a thread is the distance from a given point on one thread to the corresponding point on the next thread. Pitch is usually expressed as a fraction. Machinists sometimes make the mistake of calling the number of threads in one inch the pitch. For example, a one-inch thread has eight threads to the inch. The pitch of this thread is $\frac{1}{8}$ inch. It is wrong to say that it is an eight-pitch thread.

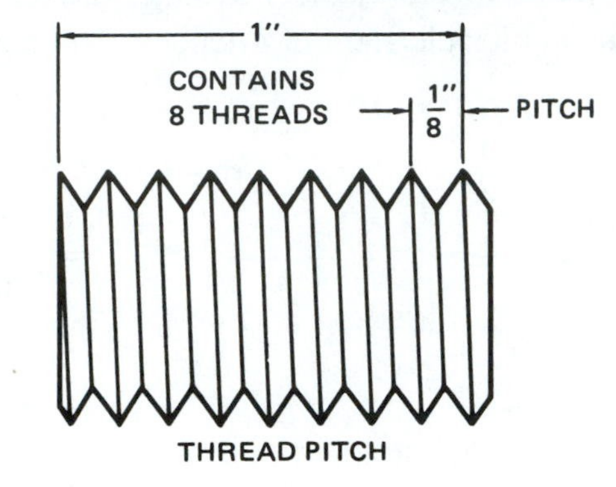

2. A screw 4 inches long has 52 threads.

   a. How many threads per inch are there? a. __________

   b. What is the pitch? b. __________

3. A screw is $3\frac{1}{2}$ inches long and has 56 threads.

   a. Find the number of threads per inch. a. __________

   b. Find the pitch of the screw. b. __________

4. On a lathe job, the tool feeds $\frac{1}{32}$ inch each time the stock turns once.

   a. How many times must the stock turn for the tool to advance $1\frac{1}{2}$ inches? a. __________

   b. How many times must the stock turn for the tool to advance $3\frac{1}{8}$ inches? b. __________

5. The feed is set for $\frac{3}{64}$ inch per revolution. How many revolutions does the stock make while the tool is advancing $6\frac{3}{8}$ inches? __________

6. A barrel of castings weighs $379\frac{1}{2}$ pounds. The empty barrel weighs 15 pounds and one of the castings weighs $6\frac{3}{4}$ pounds. How many castings are in the barrel? __________

7. A bar of steel, 22 feet 9 inches long, weighs $107\frac{11}{16}$ pounds. What does it weigh per foot? HINT: Express 22 feet 9 inches as feet. __________

8. A steel bar, $1\frac{5}{8}$ inches in diameter and 5 feet long, weighs $35\frac{1}{4}$ pounds. Find the weight of a bar of the same diameter that is 17 feet $9\frac{1}{2}$ inches long. HINT: Express 17 feet $9\frac{1}{2}$ inches as feet. __________

Note: Use this information for problems 9–12. The circumference of a circle is equal to $3\frac{1}{7}$ times the diameter. The diameter is equal to the circumference divided by $3\frac{1}{7}$.

9. The circumference of a bar of steel is $1\frac{3}{8}$ inches. What is the diameter? __________

10. The circumference of a small gear blank is $5\frac{1}{2}$ inches. Find the diameter. __________

11. What is the diameter of a small bearing that has a circumference of $\frac{11}{16}$ inch? __________

12. A ring has an outside circumference of $16\frac{1}{2}$ inches. What is its outside diameter? __________

Unit 9 COMBINED OPERATIONS WITH COMMON FRACTIONS

BASIC PRINCIPLES

- Review and apply the principles of addition, subtraction, multiplication, and division of common fractions to these problems.

PRACTICAL PROBLEMS

Note: Use this diagram for problems 1–5. Allow $\frac{3}{16}$ inch for cutting off the stock and for finishing the ends.

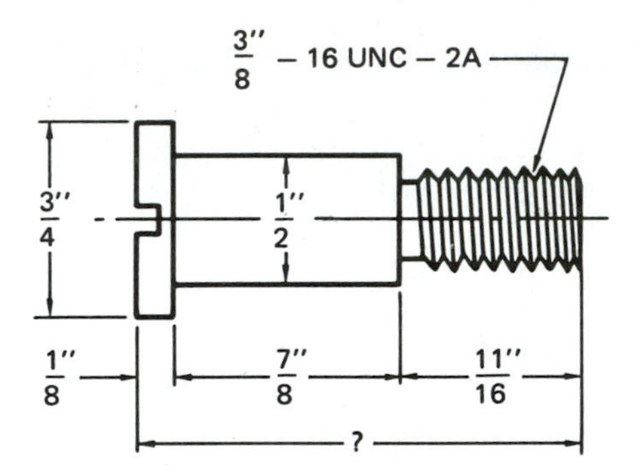

1. How many studs, like the one illustrated, can be made from a bar 1 foot 11 inches long? __________

2. How many can be cut from an 18-foot bar? __________

3. How many studs can be cut from three pieces of bar that are 6 inches, 14 inches, and 3 feet long, respectively? __________

4. In making up an order of 100 studs, two 3-foot pieces of $\frac{3}{4}$-inch stock are available in the stock rack. How many more 3-foot lengths are needed? __________

5. The stock rack contains eight pieces of 1 foot 2 inches long stock, six pieces of $4\frac{1}{2}$ feet long stock, and two bars of stock each 20 feet long. How much more stock must be ordered to fill an order for 600 studs? __________

Note: Use this diagram for problems 6–10.

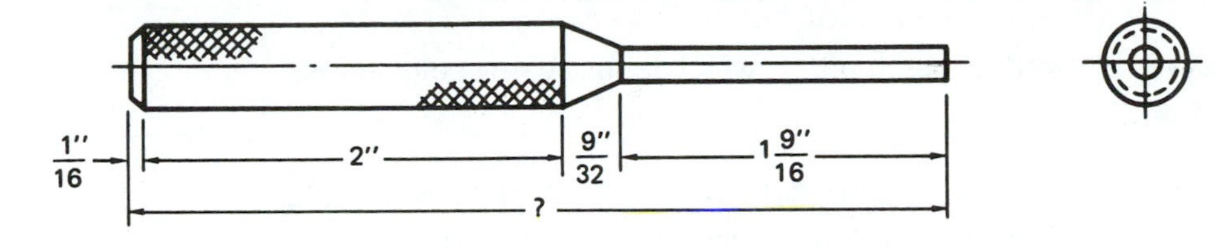

6. How long a piece of stock is required to make three of these punches? Allow $\frac{1}{16}$ inch for waste on each punch. __________
7. If $\frac{1}{8}$ inch waste is allowed on each punch, what length of stock is needed to make $\frac{1}{2}$ dozen punches? __________
8. How many punches can be made from a piece of stock 2 feet long? Allow $\frac{3}{16}$ inch waste on each punch. __________
9. Allowing $\frac{1}{8}$ inch waste on each punch, how many punches can be made from an 18-foot bar? __________
10. The available stock is: one piece, 7 inches long; two pieces, 12 inches long; and one piece, 24 inches long. Allowing $\frac{1}{8}$ inch waste for each punch, how many punches can be made? __________

Note: Use this diagram for problems 11–15.

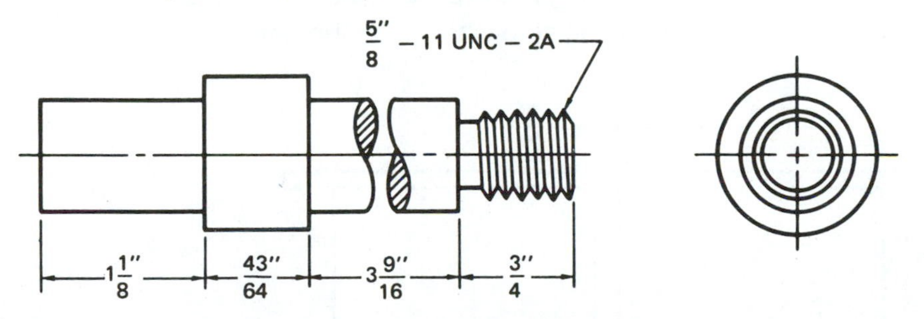

11. What length of stock is required to make this arbor? Allow $\frac{1}{8}$ inch for cutting off the stock and for finishing the ends. __________
12. What length is necessary to make three arbors, allowing $\frac{1}{8}$ inch on each for cutoff and for end finishing? __________
13. For 10 arbors, what length is sawed off a bar of stock? Allow $\frac{3}{16}$ inch on each for waste. __________

14. How many arbors can be cut from a piece of stock 3 feet $5\frac{7}{8}$ inches long? The allowance for cutting and for squaring ends is $\frac{3}{16}$ inch. ______

15. If $\frac{1}{8}$ inch waste is allowed on each arbor, which length bar results in less waste: 18 foot or 20 foot? ______

16. The weight of $1\frac{1}{8}$-inch across the flats hexagon steel bars is $3\frac{3}{4}$ pounds per running foot. Using this constant, find the weight of a $1\frac{1}{8}$-inch hexagon steel bar that is 14 feet 7 inches long. ______

CRITICAL THINKING PROBLEMS

1. Determine the full length (L) of the shaft. ______

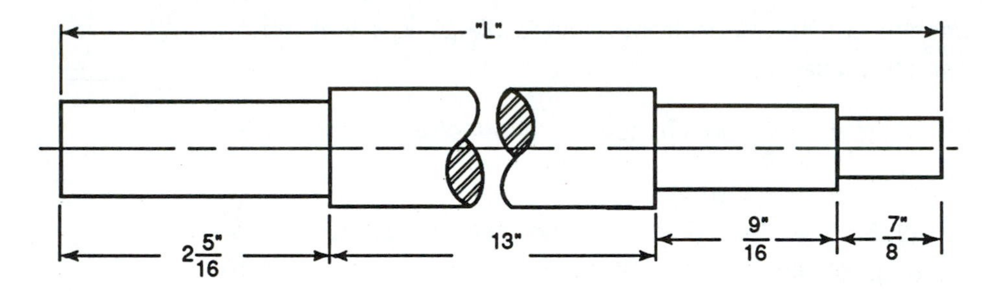

2. Determine the number of parts that can be blanked from an 8′ strip of the steel shown. $\frac{3}{16}''$ is left at the beginning of the strip and between the parts. How much remains at the end of the strip when all of the parts are blanked? ______

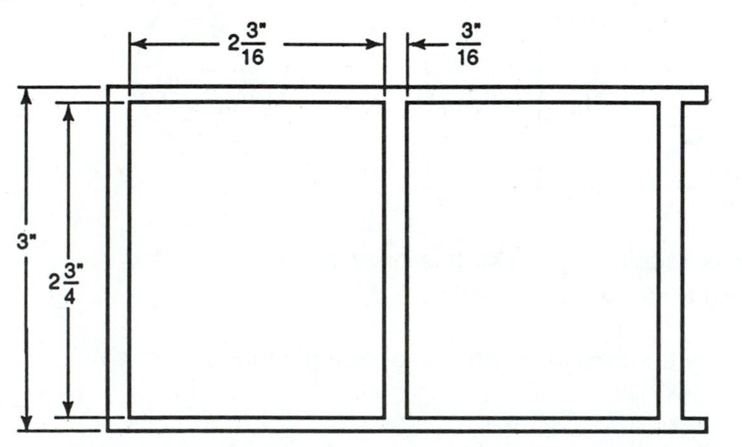

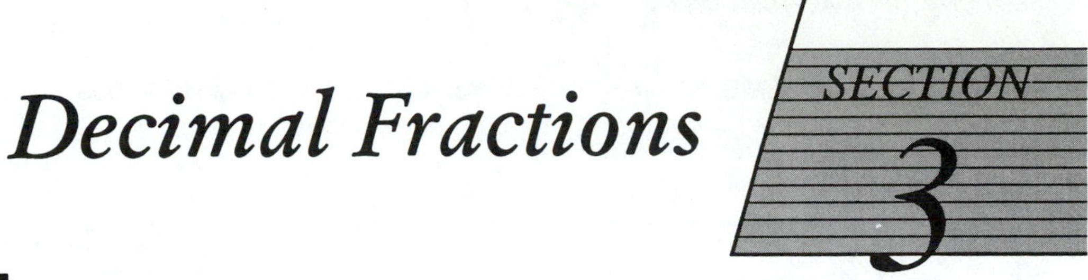

Decimal Fractions

Unit 10 ADDITION OF DECIMAL FRACTIONS

BASIC PRINCIPLES

- *Decimal fractions* are formed by dividing the numerator of a common fraction by its denominator. See *Table of Decimal Equivalents* in the appendix.

$$\frac{1}{2} = 1 \div 2 = 0.5 \text{ ans.}$$

$$\frac{7}{8} = 7 \div 8 = 0.875 \text{ ans.}$$

- Decimal fractions are added by arranging the numbers in a column with the decimals aligned. The alignment determines the position of the decimal in the answer.

Examples:

$$\begin{array}{r} 23.400 \\ +\ \ 4.567 \\ \hline 27.967 \end{array} \text{ ans.} \qquad \begin{array}{r} 3.1001 \\ 405.6000 \\ +\ \ 12.4045 \\ \hline 421.1046 \end{array} \text{ ans.}$$

REVIEW PROBLEMS

a.
$$\begin{array}{r} 1.2 \\ +\ 3.2 \\ \hline \end{array}$$

b.
$$\begin{array}{r} 67.450 \\ 4.789 \\ +\ 123.500 \\ \hline \end{array}$$

c.
$$\begin{array}{r} \$\ 4,567.98 \\ 343.87 \\ +\ \ 19,999.34 \\ \hline \end{array}$$

PRACTICAL PROBLEMS

Note: Use this diagram for problems 1–4.

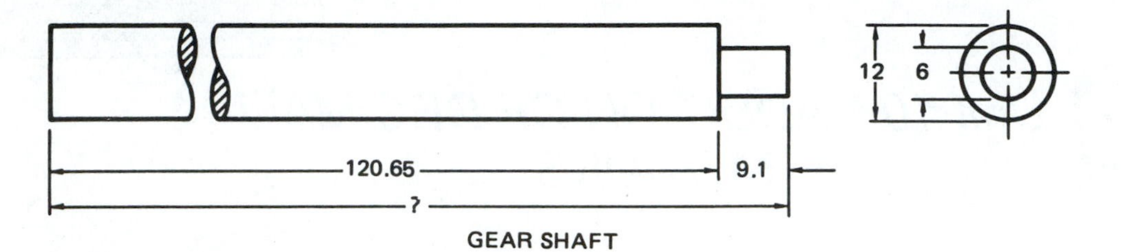

GEAR SHAFT

1. What is the overall length of the gear shaft in millimeters? __________

2. The dimensions 120.65 mm and 9.1 mm are changed to 105.72 mm and 79.75 mm, respectively. What is the overall length of the shaft? __________

3. If the dimensions are 6.3125 inches and 0.5625 inches, what is the total length? __________

4. The dimensions 101.630 mm and 94.708 mm replace the dimensions of length for the gear shaft. What is the overall length of the shaft? __________

Note: Use this diagram for problems 5–7.

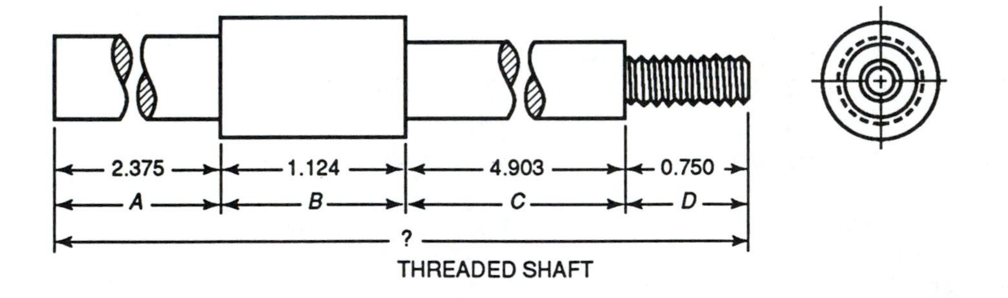

THREADED SHAFT

5. Find, in inches, the length of the threaded shaft without the threaded part. __________

6. Find the overall length of the shaft using these dimensions for ***A***, ***B***, ***C***, and ***D***. ***A*** = 94.5 mm; ***B*** = 27.7 mm; ***C*** = 271.9 mm; ***D*** = 20 mm. __________

7. If ***A*** = 47.8 mm, ***B*** = 20.03 mm, ***C*** = 319.7 mm, and ***D*** = 70.7 mm, what is the overall length of the shaft? __________

Note: Use this diagram for problems 8–13.

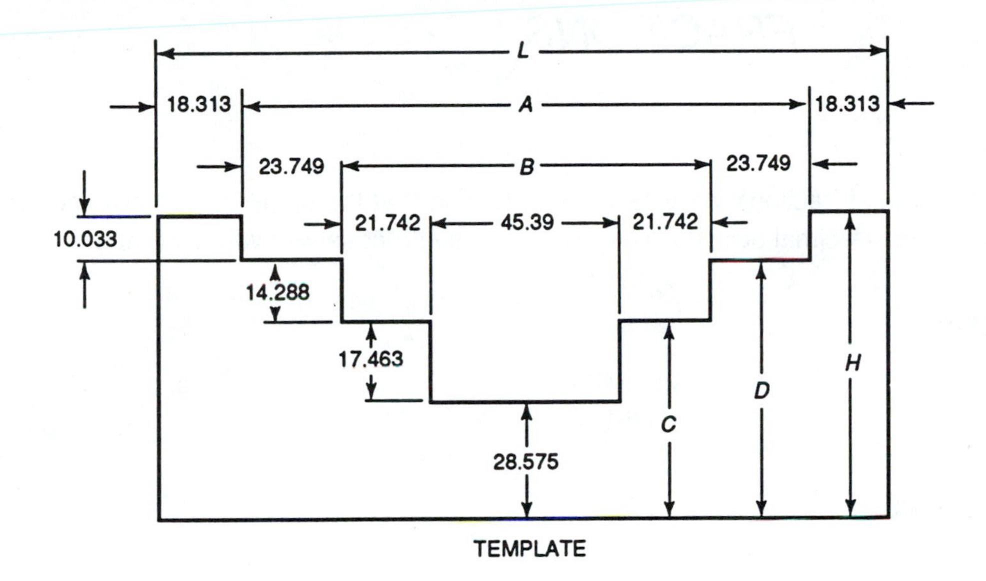

TEMPLATE

8. What is the total length of the template (dimension ***L***) in millimeters? ____________

9. What is the total height ***H*** in millimeters? ____________

10. Find, in millimeters, dimension ***A***. ____________

11. What is dimension ***B*** in millimeters? ____________

12. Find, in millimeters, dimension ***C***. ____________

13. What is dimension ***D*** in millimeters? ____________

Unit 11 SUBTRACTION OF DECIMAL FRACTIONS

BASIC PRINCIPLES

- To subtract decimal fractions, arrange the problem so that the smaller number is below the larger and the decimal points are aligned. Then subtract as with whole numbers.

REVIEW PROBLEMS

a. 56.9
 \- 3.6

b. 9883.456
 \- 298.179

c. $456.34
 \- 331.56

PRACTICAL PROBLEMS

Note: Use this diagram for problems 1–3.

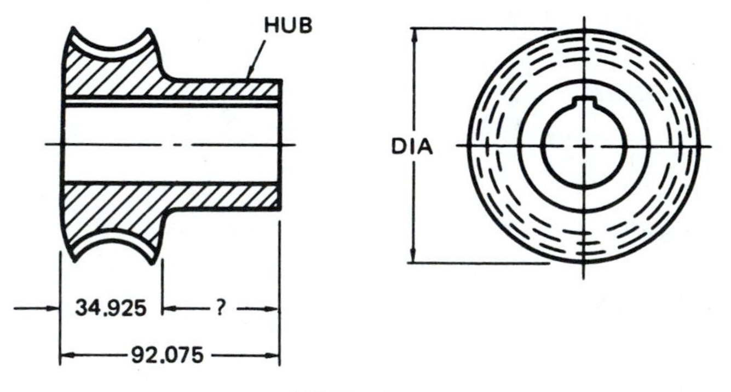

WORM GEAR

1. What is the length of the hub on this worm gear in millimeters? __________

2. If the overall length is 114.3 mm and the hub length is 68.262,5 mm, what is the gear thickness? __________

3. If the overall length is 107.95 mm and the gear thickness is 32.131 mm, what is the hub length? __________

Note: Use this information and diagram for problems 4–7. If a piece changes in diameter or width at a constant rate for a part of its length, that part is called a *taper.* The total amount of taper, often called the *total taper,* is the difference in the diameters of the large and small ends of the taper.

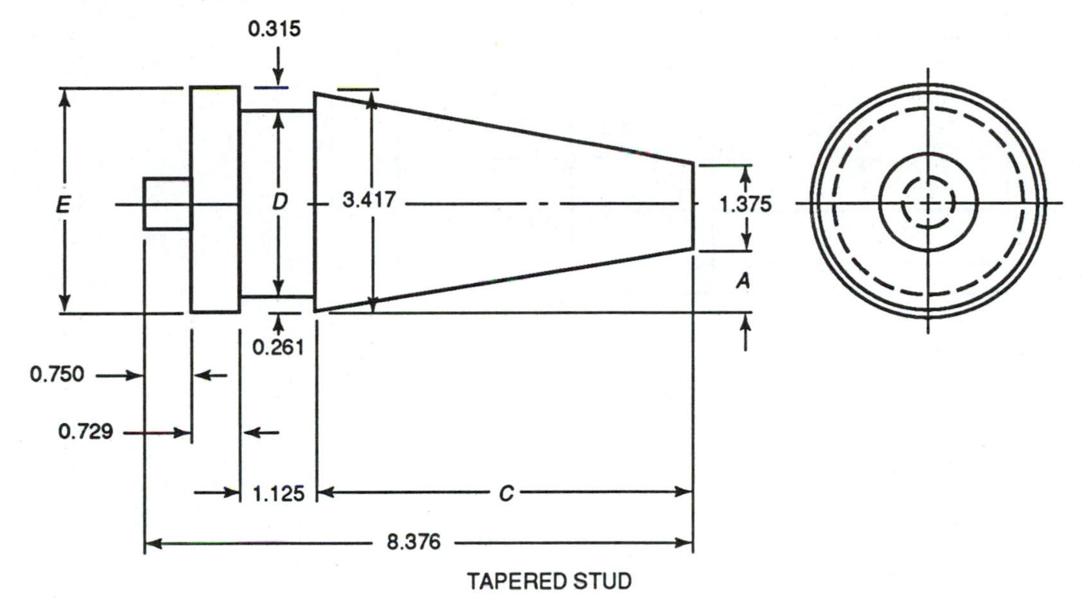

TAPERED STUD

4. What is the total taper of the tapered stud in inches? __________

5. Find, in inches, dimension ***C***. __________

6. What is dimension ***D*** in inches? __________

7. Find, in inches, dimension ***E***. __________

8. The outside diameter of a bushing is 61.443 mm and the wall thickness is 9.525 mm. What is the inside diameter? __________

9. What is the inside diameter of a bushing that has an outside diameter of 61.417 mm and a wall thickness of 22.225 mm? __________

10. A lathe operator turns a 0.094-inch chip from a bar 3.952 inches in diameter. What is the finished diameter of the bar? __________

11. The diameter of a bar is 69.85 mm before turning. The depth of the cut is 1.702 mm. Find the finished diameter of the bar. __________

12. This gauge drawing is sent to a tool for use in making the gauge. The supervisor finds that three important dimensions are missing.

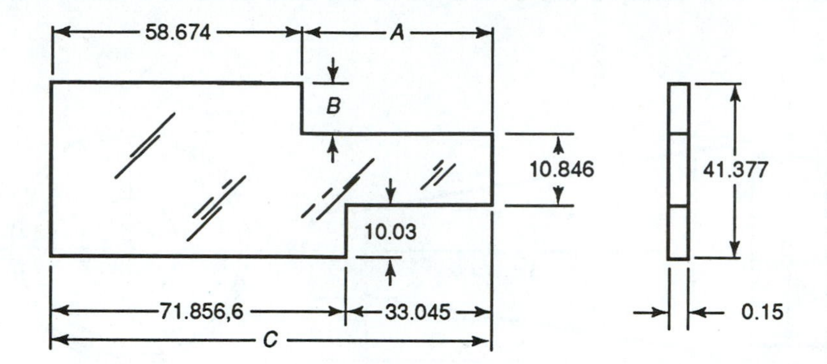

a. Find, in millimeters, dimension ***C***. a. ________

b. What is dimension ***B*** in millimeters? b. ________

c. Find, in millimeters, dimension ***A***. c. ________

Unit 12 MULTIPLICATION OF DECIMAL FRACTIONS

BASIC PRINCIPLES

- Multiplication of decimal fractions is the same as with whole numbers except for the placement of the decimal in the answer. To place the decimal in the answer, count the number of decimal places to the right of the decimal in both numbers being multiplied, add the places, and then count over from the right and place the decimal in the answer.

Example:

3.4567	⟵ 4 decimal places
x 3.9876	⟵ 4 decimal places
13.78393692	⟵ 8 decimal places

PRACTICAL PROBLEMS

Note: Use this diagram for problems 1–3.

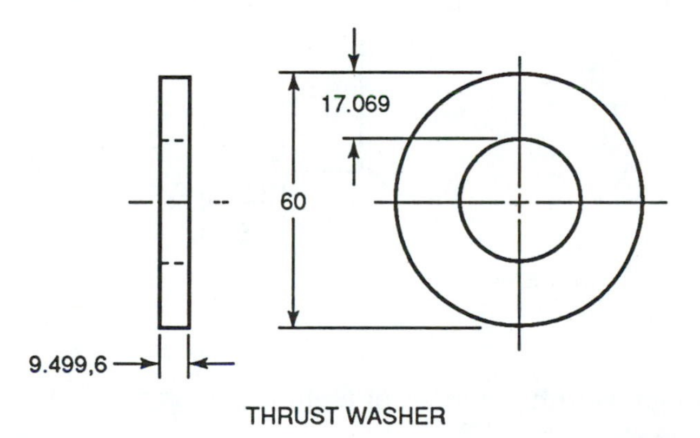

THRUST WASHER

1. How many millimeters will three of the illustrated thrust washers take up on a shaft? __________

2. How many millimeters high is a stack of 13 of these washers? __________

3. Find, in millimeters, the diameter of the hole in this thrust washer. __________

Note: Use this information for problems 4–6. To find the circumference of a circle, multiply the diameter by 3.1416. This is written:

$$\textit{circumference} = 3.1416 \times \textit{diameter}$$

or

$$C = 3.1416 \times D$$

4. The diameter of a circle is 65 mm. Find the circumference. __________

5. If the diameter is 48.5 millimeters, what is the circumference? __________

6. Find the circumference of a 3.267-inch diameter circle. Round the answer to four decimal places. __________

Note: Use this information for problems 7–10. The *circular pitch* of a gear is the distance along the pitch circle between the centers or other corresponding points of the adjacent teeth. To find the pitch circumference, multiply the circular pitch by the number of teeth.

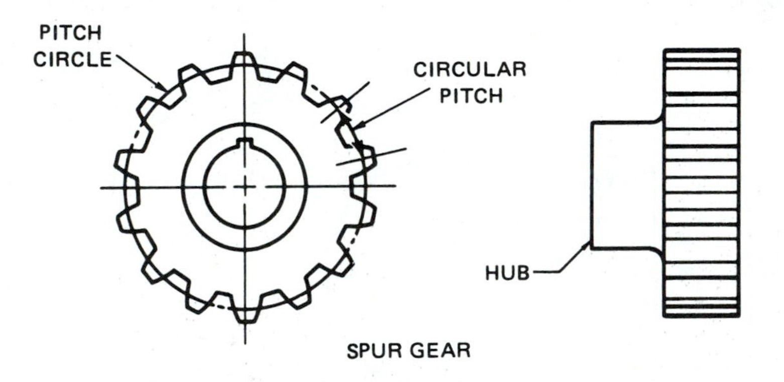

SPUR GEAR

7. The circular pitch is 0.095 inch and the number of teeth is 20, find the circumference of the pitch circle. __________

8. If circular pitch is 0.287 inch and the number of teeth is 42, find the circumference of the pitch circle. __________

9. The circular pitch is 1.5904 inches. The number of teeth is 125. What is the circumference of the pitch circle? __________

10. If the circular pitch is 0.9382 inch and the number of teeth is 51, find the circumference of the pitch circle. __________

11. A cast iron pulley contains 13.6 cubic inches of metal. It weighs 0.26 pound per cubic inch. What is the weight of the pulley? ____________

12. The distance across the corners of a square is always equal to the length of one side times 1.414. Determine the diameter of a round hole that will allow the square shaft shown to pass through it. Round the answer to four decimal places. ____________

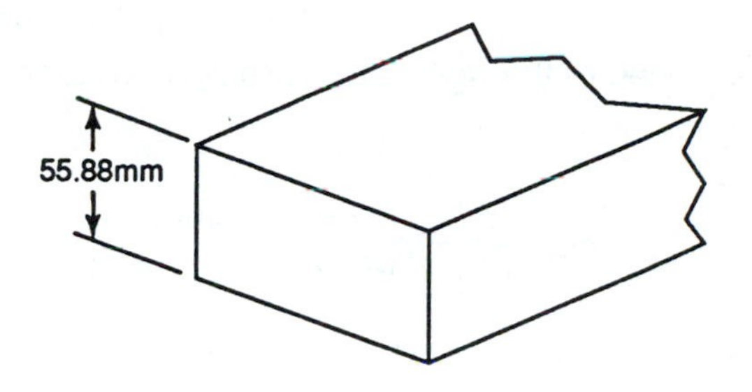

Unit 13 DIVISION OF DECIMAL FRACTIONS

BASIC PRINCIPLES

- Dividing decimal fractions requires moving the decimal point of the divisor to the extreme right to make a whole number divisor. The decimal in the dividend is then moved the same number of places to the right, and zeroes are added to the dividend if it has less digits than the divisor.

```
                          5.5   ←——   Quotient
Divisor ——→   5.35. ) 29.42.5   ←——   Dividend
```

The decimal position in the quotient will now be directly above the decimal in the dividend.

Finally, divide as in whole number division, placing additional zeroes in the dividend if it cannot be divided evenly by the divisor. Note that each number in the quotient is placed directly above the number involved in the dividend.

```
                5.5   ans.
   5 35. ) 29 42.5
           26 75
            2 67 5
            2 67 5
                 0
```

- Many times it will be necessary to round off the decimal remainder of your answer. For example, if the answer to a money problem has three or more digits to the right of the decimal point, we round the number off to the hundredths place (second digit to the right of the decimal). This is the cents position. A machinist will generally round numbers to the thousandths (third digit to the right of the decimal point), or to the ten thousandths position (fourth digit to the right of the decimal point), because measuring and machine tools used are usually graduated in these units.

- The rules used in this text for rounding numbers follow:

  1. If the number to the right of the last significant digit is equal to or greater than five, increase the last significant digit by one and drop all digits to the right of the last significant digit.

2. If the digit to the right of the last significant digit is less than five, leave the last significant digit as it is and drop all digits to the right of the last significant digit.

Examples:

Number	ROUNDED TO: 4 digits	3 digits	2 digits
3.1416	3.142	3.14	3.1
14.815	14.82	14.8	15
321.35	321.4	321	320
6,274.5	6,275	6,270	6,300
20,018	20,020	20,000	20,000
71,853	71,850	71,900	72,000

PRACTICAL PROBLEMS:

1. Occasionally in manufacturing it must be determined how many threads will be cut in a specific length of work. The number of threads per inch or per millimeter is $n = \frac{1}{p}$. Complete this chart for the number of threads in each given length of work.

	Pitch	Length	Threads Per Inch or Threads Per Millimeter	Number of Threads in Total Length
a.	0.125 in.	$2\frac{1}{2}$ in.		
b.	0.0625 in.	1.875 in.		
c.	0.05 in.	$12\frac{3}{4}$ in.		
d.	0.03125 in.	$3\frac{5}{8}$ in.		
e.	0.25 mm	20 mm		
f.	1.6 mm	104 mm		
g.	1.25 mm	47.5 mm		
h.	3.2 mm	57.6 mm		

2. Cold rolled steel, 1 inch in diameter, weighs 2.68 pounds per foot of length. How many feet are there in a bundle of bars weighing 203.5 pounds? Express the answer to the nearer hundredth. __________

3. One liter of machine oil weighs 0.941 kilogram. A tank partly filled with this oil weighs 58.287 kilograms. The tank itself weighs 7.257 kilograms. Find, to the nearer thousandth, how many liters of oil there are in the tank. __________

Note: Use this information for problems 4 and 5. In spur gears, the depth of the cut which forms the teeth of the gear is equal to 2.250 divided by the diametral pitch of the gear. ($h_t = 2.250 \div P$). This depth is expressed in inches or in millimeters and is usually expressed to the nearer ten-thousandth. h_t = depth of cut. P = diametral pitch.

4. What is the depth of the teeth in a set of lathe gears that have a diametral pitch of 16 inches? Express the answer to the nearer ten-thousandth inch. __________

5. A gear with a 12-inch diametral pitch is used on a milling machine attachment. What is the depth of the teeth on the gear? Express the answer to the nearer ten-thousandth inch. __________

Note: Use this diagram for problem 6.

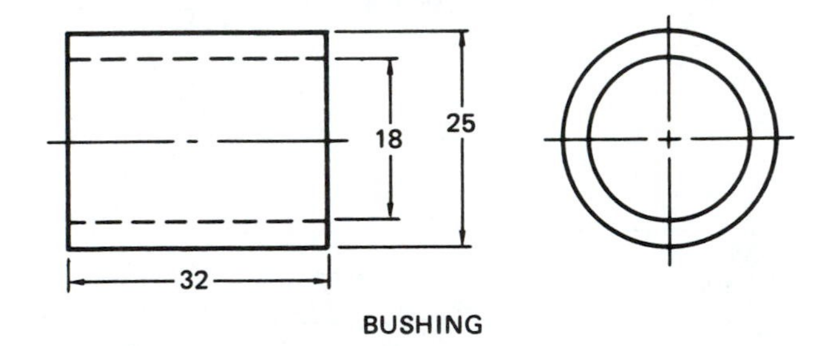

BUSHING

6. Bushings are cut from a bronze bar 362.5 mm long. There are 1.5 mm of waste on each piece.

   a. How many bushings can be cut from this piece? a. __________

   b. How long is the piece that is left? b. __________

7. The circumference of a pulley is equal to 3.1416 times the diameter. The diameter is equal to the circumference divided by 3.1416. Find, to the nearer ten-thousandth, the diameter of each pulley.

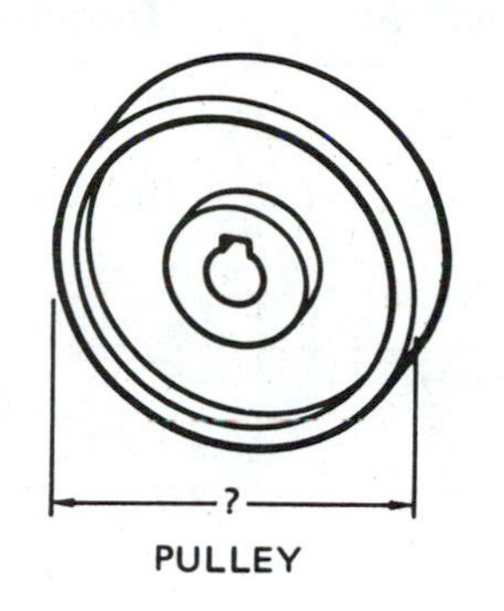

	Circumference	Diameter
a.	5.000 in.	
b.	13.750 in.	
c.	21.500 in.	
d.	912.5 mm	
e.	692.5 mm	
f.	127.05 mm	

8. Express each fraction as a decimal.

	Fraction	Decimal
a.	$\frac{5}{8}$	
b.	$\frac{3}{4}$	
c.	$\frac{7}{16}$	
d.	$\frac{11}{16}$	
e.	$\frac{31}{32}$	
f.	$\frac{17}{64}$	

	Fraction	Decimal
g.	$\frac{23}{64}$	
h.	$\frac{53}{64}$	
i.	$\frac{19}{32}$	
j.	$\frac{63}{64}$	
k.	$\frac{27}{32}$	
l.	$\frac{47}{64}$	

9. Express each fraction as a decimal. Round each answer to the nearer ten-thousandth.

	Fraction	Decimal
a.	$\frac{13}{17}$	
b.	$\frac{7}{9}$	
c.	$\frac{7}{12}$	

	Fraction	Decimal
d.	$\frac{11}{13}$	
e.	$\frac{3}{11}$	
f.	$\frac{127}{147}$	

10. Express 0.7 inch as a fraction with a denominator of 32. ____________

11. Express 0.31 inch as a fraction with a denominator of 64. ____________

12. Express 0.566 inch to the nearer $\frac{1}{16}$ inch. ____________

13. Express 0.86 inch to the nearer $\frac{1}{32}$ inch. ____________

14. Express 0.809 inch as a fraction with a denominator of 16. ____________

Unit 14 COMBINED OPERATIONS WITH DECIMAL FRACTIONS

BASIC PRINCIPLES

- Review and apply the principles of *addition, subtraction, multiplication,* and *division of decimal fractions* to these problems.

PRACTICAL PROBLEMS

1. What is the total length of this clevis in inches? ____________

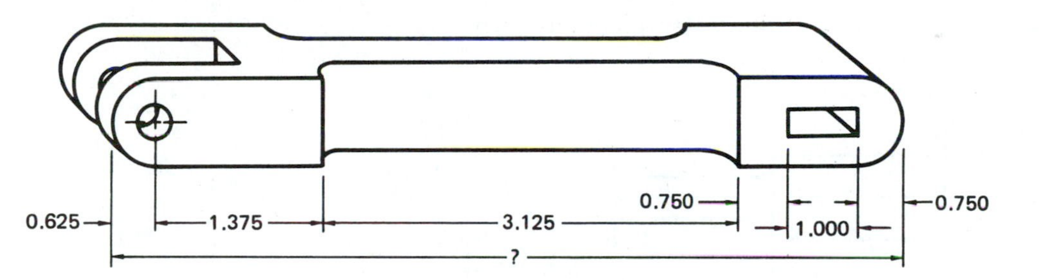

2. Three strips of brass with thickness of 0.53 mm, 1.47 mm, and 0.12 mm are placed on top of each other. What is the combined thickness of the strips? ____________

3. A profile gauge is shown.

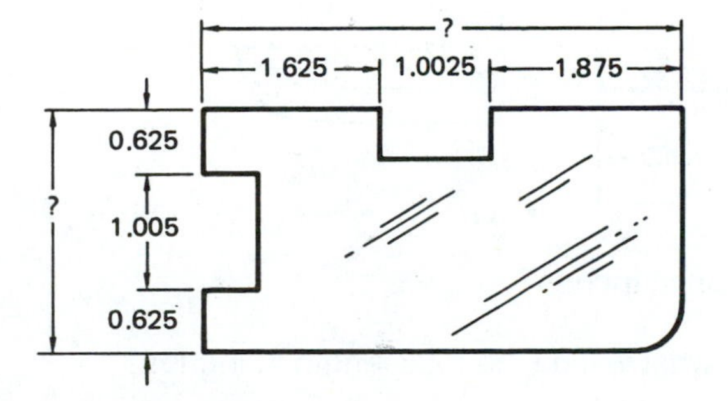

   a. Find, in inches, the length of the gauge. a. ____________

   b. Find, in inches, the width of the gauge. b. ____________

4. Determine, in millimeters, the outside diameter of this steel tube. __________

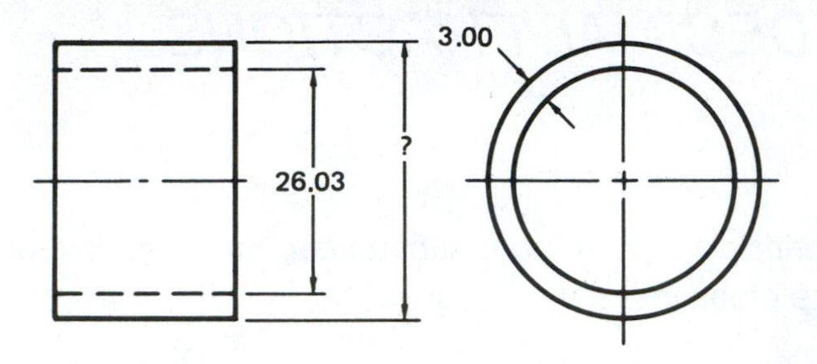

5. The inside dimension of a gas pipe is 12.52 mm. The thickness of the metal is 2.31 mm. What is the outside diameter? __________

6. Determine, in millimeters, the total length of this crank pin. __________

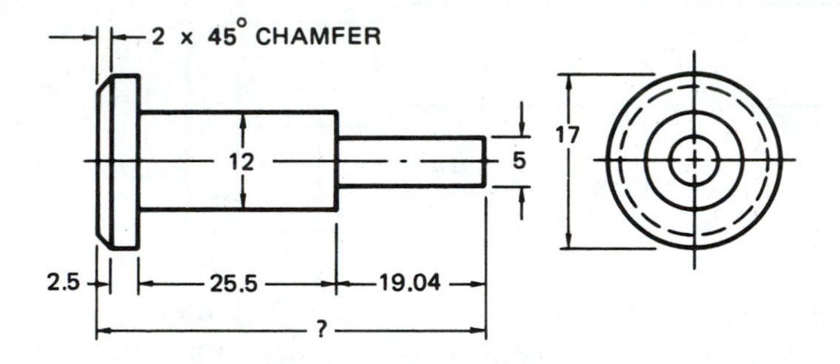

Note: Use this diagram for problems 7–10.

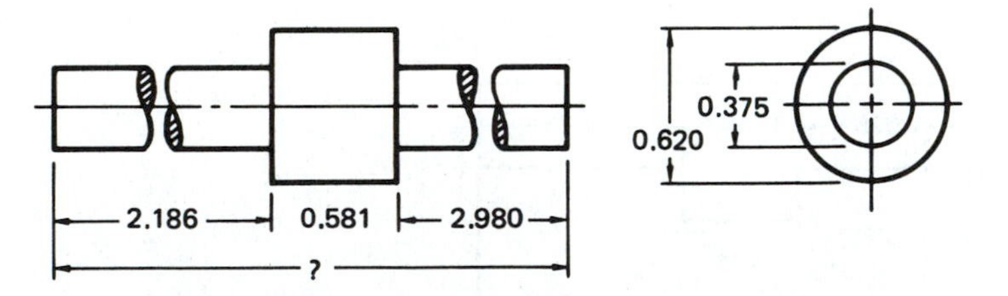

7. What is the total length of this pin in inches? __________

8. If the pin is enlarged 2.5 times, what will be the total length in inches? __________

9. If the pin is enlarged 2.5 times, what will be the outside diameter in inches? __________

10. If the pin is enlarged 2.5 times, what will be the inside diameter in inches? __________

11. A piece of $1\frac{1}{2}$-inch diameter *1020* steel is turned down to a 1.037-inch diameter. How deep is the tool fed into the work by the lathe cross slide? ________

12. What is the length of the threaded part of the bolt in inches? ________

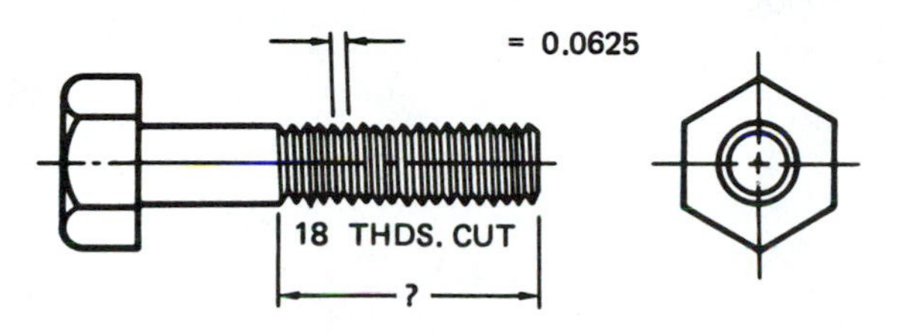

13. The hole in a sleeve is bored out for a 2"-$4\frac{1}{2}$ UNC thread. The thread depth on each side is 0.120 in. What is the diameter of the bored hole? ________

14. The minor diameter of a $1\frac{1}{2}$-inch *Unified National Coarse* threaded bolt is 1.2955 inches. What is the thread depth? ________

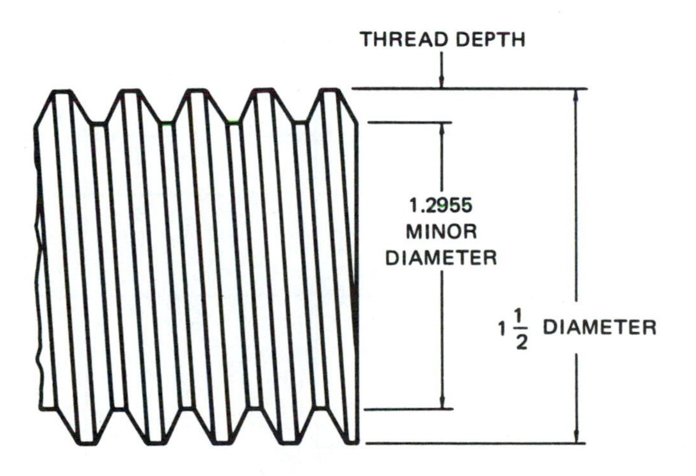

15. One pound of number *10* copper wire measures 31.82 feet. How many pounds does a coil containing 125 feet of this wire weigh? Round the answer to the nearer ten-thousandth pound. ________

CRITICAL THINKING PROBLEMS

1. A drawing indicates that a 3.5625" diameter bored hole has a tolerance of +0.0025", -0.001". Determine the maximum and minimum dimensions that the hole can be and still be a good part.

2. DETERMINE DIMENSIONS:

   A __________

   B __________

   C __________

   D __________

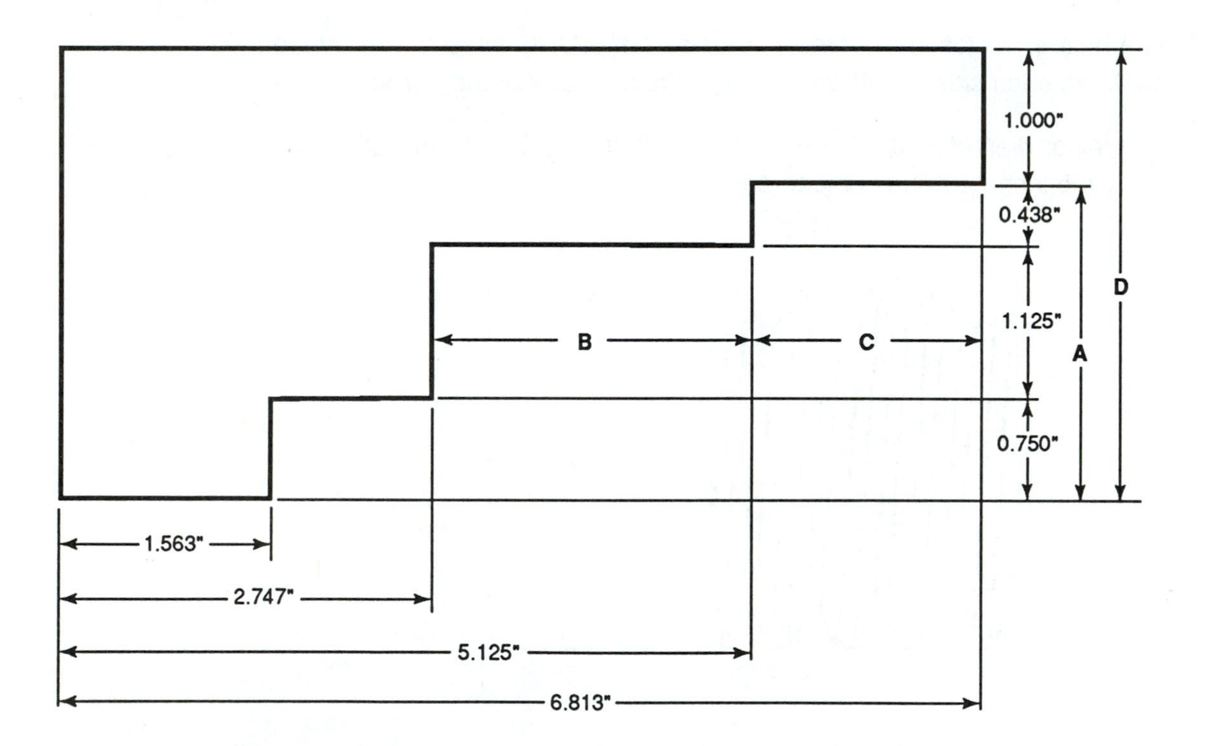

Direct Measure

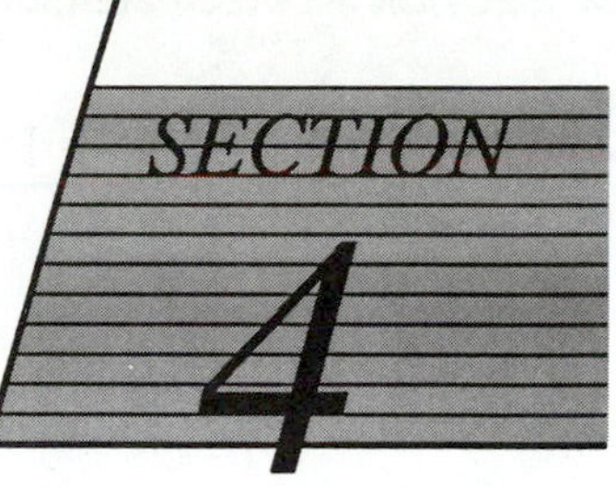

Unit 15 LENGTH MEASURING INSTRUMENTS

BASIC PRINCIPLES

- One of the most important skills a machinist can have is the ability to read measuring instruments quickly and accurately. There are many new electronic instruments available, but the older vernier devices are still used in many shops.

- The most common measuring tools used in the machine shop are the *steel scale,* the *vernier caliper,* and the *micrometer.* Steel scales come in many combinations of

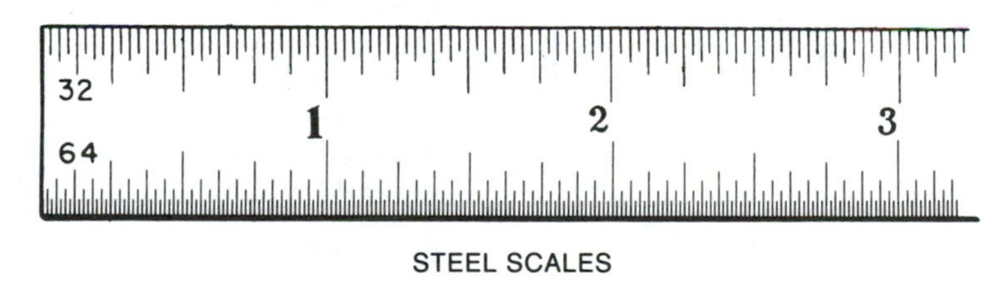

STEEL SCALES

Courtesy of Olivo and Olivo: BASIC TECHNICAL MATHEMATICS, 6th edition, ©1992 by Delmar Publishers Inc.

graduations, but generally they are not used for precise work that requires measurements smaller than $\frac{1}{64}$ of an inch. Vernier calipers can be used for inside and outside measurements and generally are not used for measurements smaller than 0.001 inch. Micrometers are probably the most used and most accurate measuring tools found in the machine shop today. Micrometers are easily accurate to 0.001 of an inch, and many have vernier graduations which increase their accuracy to 0.0001 of an inch.

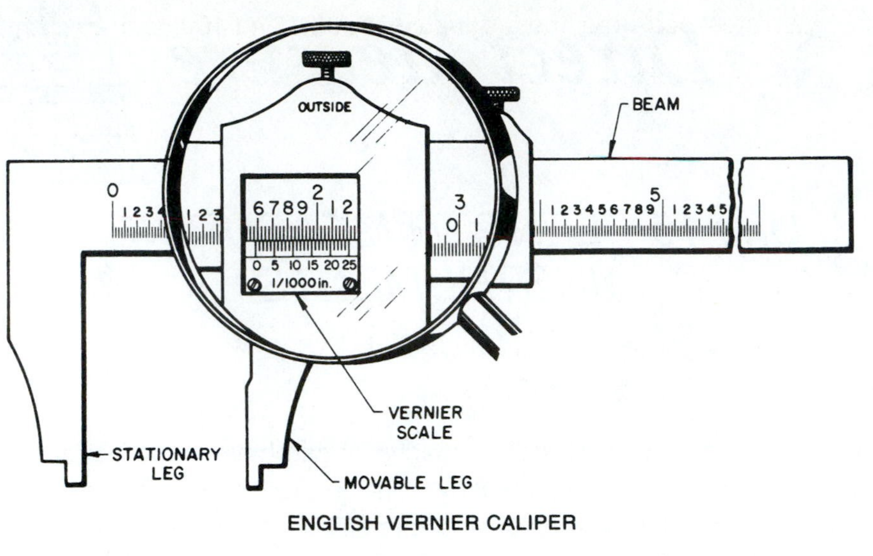

ENGLISH VERNIER CALIPER

Courtesy of Olivo and Olivo: BASIC TECHNICAL MATHEMATICS,
6th edition, ©1992 by Delmar Publishers Inc.

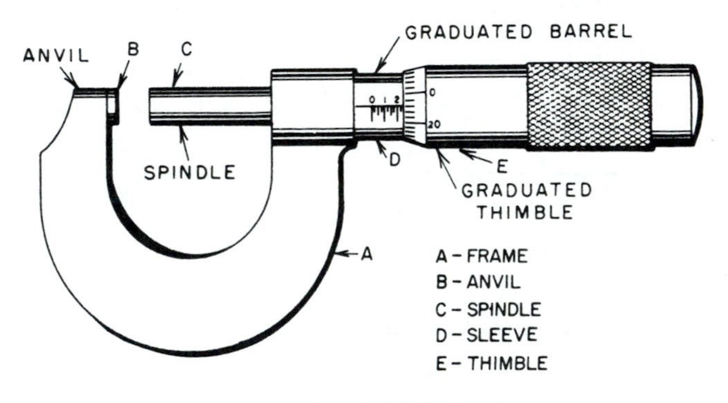

ENGLISH MICROMETER

Courtesy of Olivo and Olivo: BASIC TECHNICAL MATHEMATICS,
6th edition, ©1992 by Delmar Publishers Inc.

- In the last few years metric instruments have become more common in the United States, so it is has become necessary to be able to read metric instruments as well. It must be noted that all of the measuring instruments discussed come in metric versions also.

- An English steel scale may be graduated in 50ths and 100ths of an inch.

Example: Determine the value of the measurements shown on the English scale.

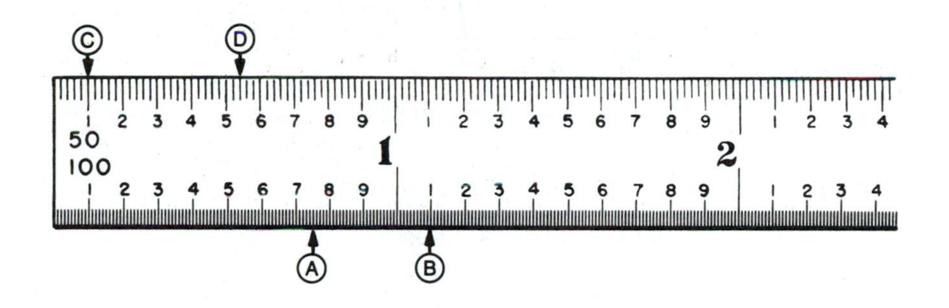

Courtesy of Olivo and Olivo: BASIC TECHNICAL MATHEMATICS,
6th edition, ©1992 by Delmar Publishers Inc.

Measurement "A" is $^{75}/_{100}''$ and would generally be reported in the simplified form of $^{3}/_{4}''$. In some cases it would be converted to a decimal fraction and reported as 0.750″. Measurement "B" is $1^{10}/_{100}''$ and would be reported as $1^{1}/_{10}''$ or 1.10″. Measurement "C" is $^{5}/_{50}''$. In simplified form it would be reported as $^{1}/_{10}''$ or 0.10″. Measurement "D" is $^{27}/_{50}''$ and could be reported simply as $^{27}/_{50}''$ or as 0.540″ since each small division is equal to 0.020″.

- Reading vernier caliper scales requires a bit of practice. The beam is graduated in 0.025″ increments, with major graduations at 0.100″ and whole inches.

  To determine the readings shown on the English vernier calipers:

  Step 1: Note the reading on the beam that is to the left of the index mark, (zero) on the vernier scale.

  Step 2: Look along the vernier scale until one of the marks on the vernier scale lines up with one of the marks on the beam. Read the value on the vernier scale and add it to the reading taken in Step 1. The graduations on the vernier scale are in 0.001".

Example: Determine the values shown on the English vernier calipers.

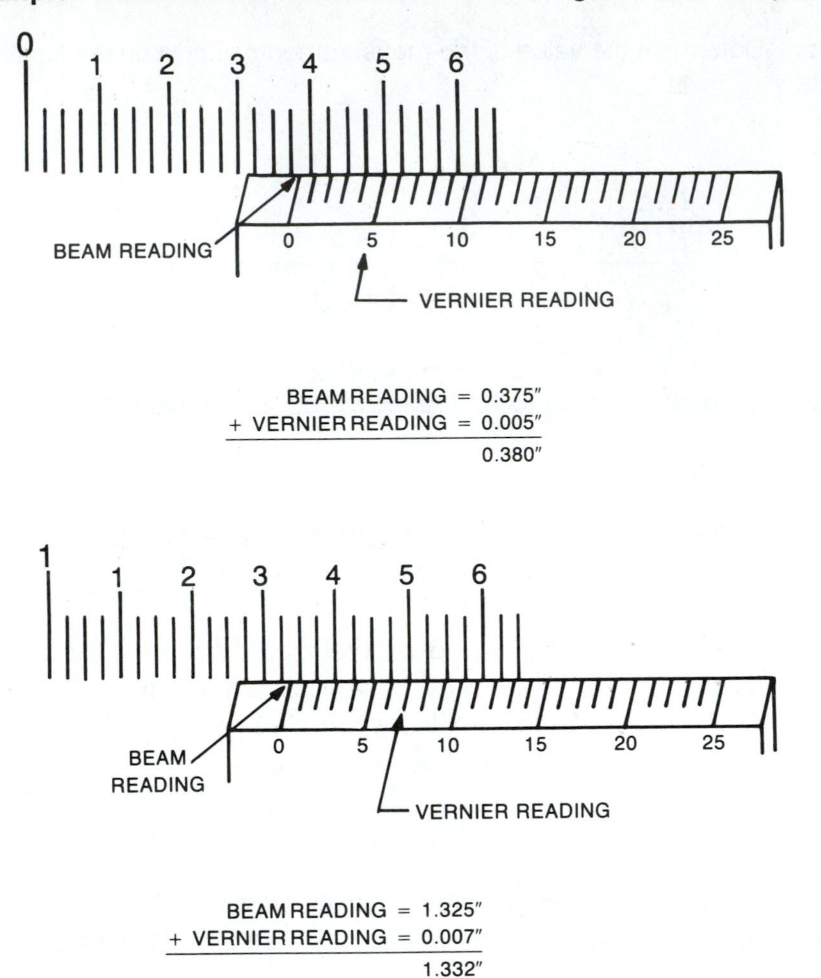

The metric vernier caliper is read the same as the English version with the exceptio that the beam is graduated in millimeters and the vernier is graduated in 0.020 mr increments. Some have both an internal and an external scale on them so be sure t read the scale appropriate to the type of reading being made.

Example: Determine the value shown on the metric vernier caliper.

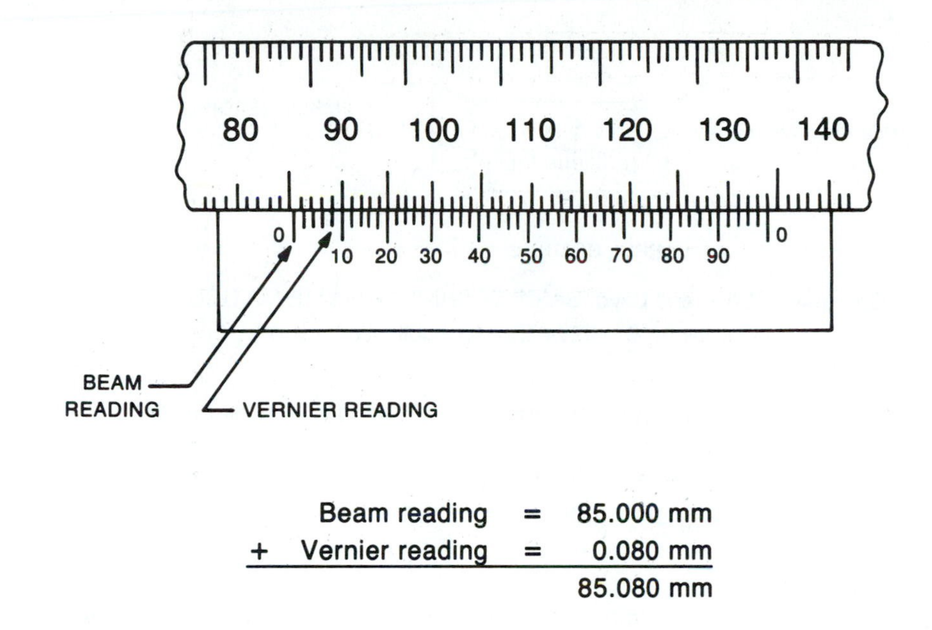

Beam reading	=	85.000 mm
+ Vernier reading	=	0.080 mm
		85.080 mm

- Micrometers are among the most common precision measuring tools found in the machine shop. The micrometer is generally used to measure to an accuracy of 0.001". Many are capable of measuring to an accuracy of 0.0001" with the addition of a vernier scale.

- The spindle threads have a pitch of 0.025"; therefore one turn of the thimble will move the thimble 0.025". The small graduations on the barrel of the micrometer are 0.025" increments. The thimble is graduated around its circumference with 25 evenly spaced marks. If the thimble is turned one graduation, the thimble will move 0.001". A summary of the graduations on the English micrometer follows:

   a. Large numbered graduations on the barrel are at 0.100" increments.

   b. Small graduations on the barrel are at 0.025" increments.

   c. Each graduation on the thimble of the micrometer is 0.001".

   d. Graduations on the back side of the barrel are the vernier scale and are at increments of 0.0001".

Example: Determine the value of the measurements on the following micrometer.

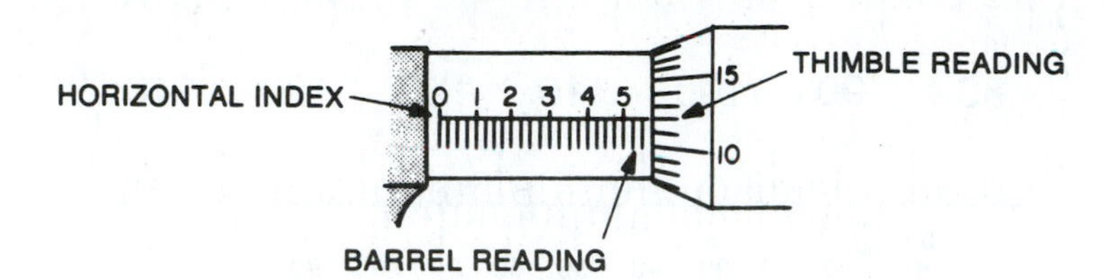

Courtesy of Olivo and Olivo: BASIC TECHNICAL MATHEMATICS,
6th edition, ©1992 by Delmar Publishers Inc.

The horizontal line on the barrel is the index mark used when reading the micrometer.

Barrel reading	=	0.550"
Thimble reading	=	0.012"
		0.562"

Example: Determine the value of the measurement on the following vernier micrometer.

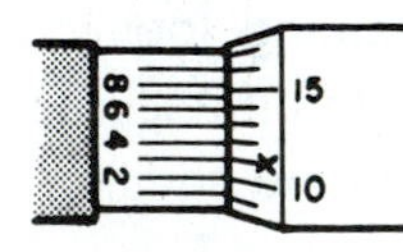

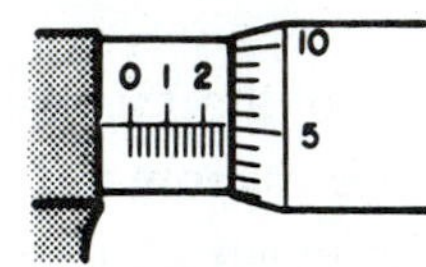

Barrel reading	=	0.250″
Thimble reading	=	0.005″
Vernier reading	=	0.0003″
		0.2553″

Courtesy of Olivo and Olivo: BASIC TECHNICAL MATHEMATICS,
6th edition, ©1992 by Delmar Publishers Inc.

Reading the metric vernier micrometer is similar to reading an English micrometer except:

a. Notice that the barrel has graduations above and below the index line. The graduations above the line are in 1 mm increments; the graduations below the line are in 0.5 mm increments.

b. When taking a barrel reading, first read the whole millimeters above the line. Then add 0.5 mm for every visible graduation line to the right of the initial barrel reading.

c. Thimble graduations are in 0.1 mm increments; vernier graduations are in 0.001 mm increments. Both of these must then be added to the barrel reading.

Example: Determine the value of the measurement shown on the following metric vernier micrometer.

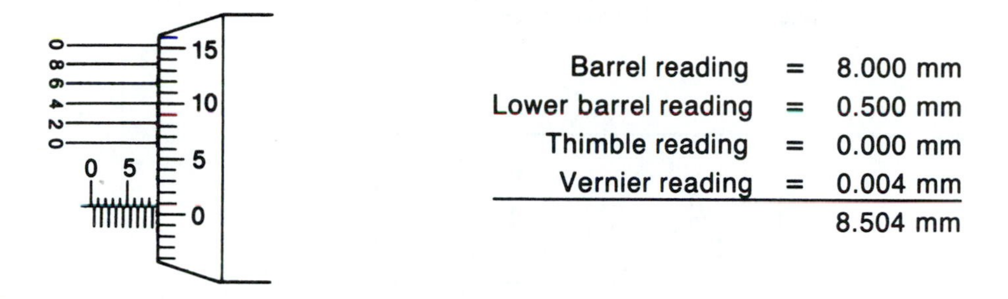

Courtesy of Olivo and Olivo: BASIC TECHNICAL MATHEMATICS, 6th edition, ©1992 by Delmar Publishers Inc.

PRACTICAL PROBLEMS

1. Determine the value of each measurement.

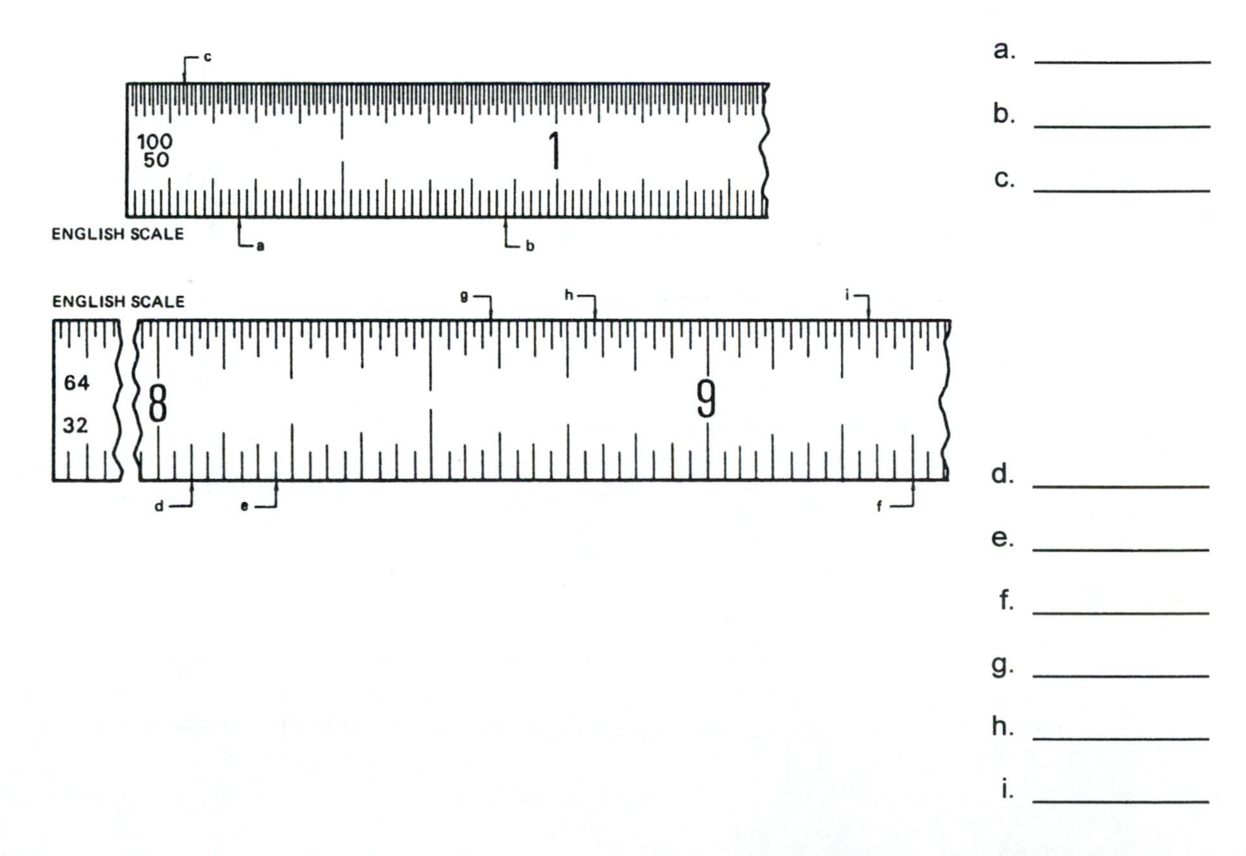

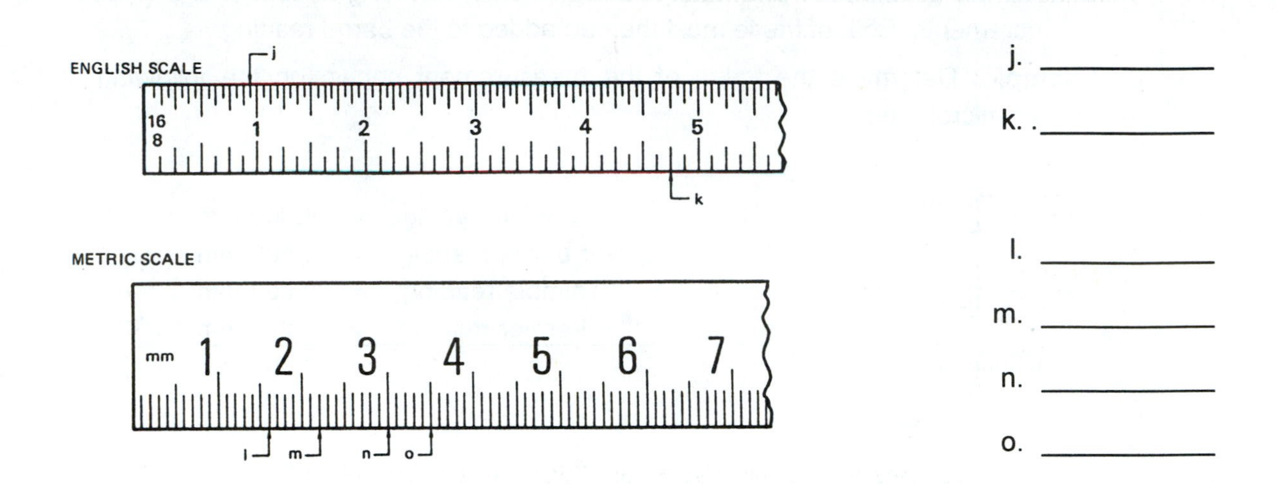

j. ________

k. ________

l. ________

m. ________

n. ________

o. ________

2. Measure the length of each line to the indicated degree of precision.

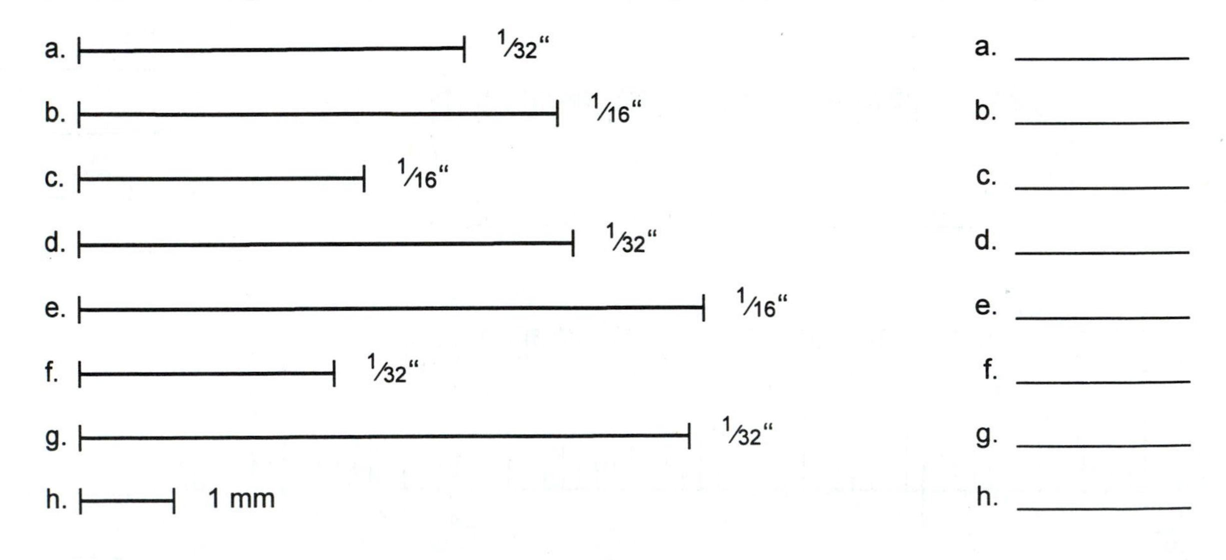

a. ________

b. ________

c. ________

d. ________

e. ________

f. ________

g. ________

h. ________

3. Determine the value of each micrometer reading.

a.

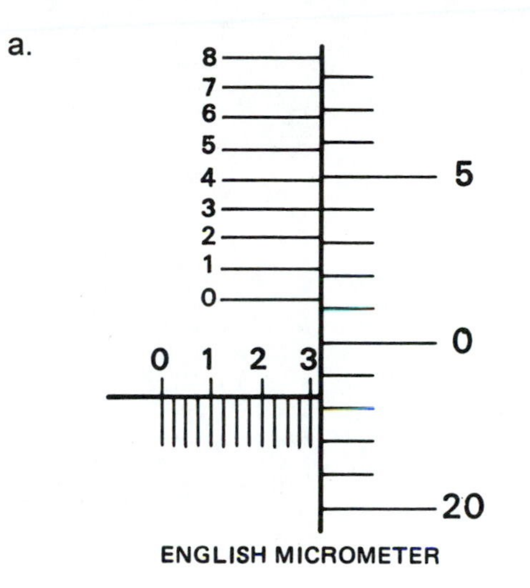

ENGLISH MICROMETER

a. ____________

b.

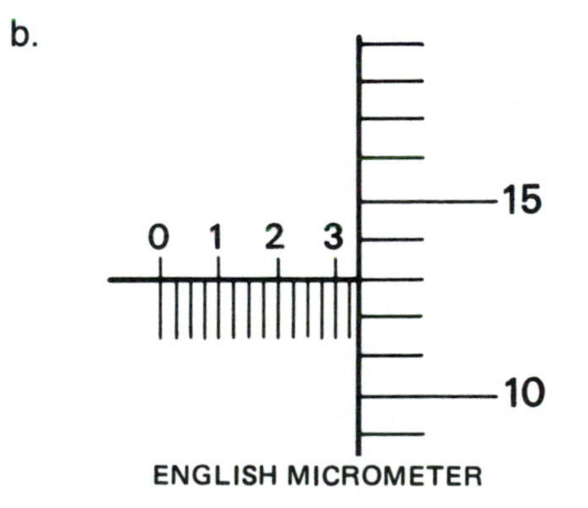

ENGLISH MICROMETER

b. ____________

c.

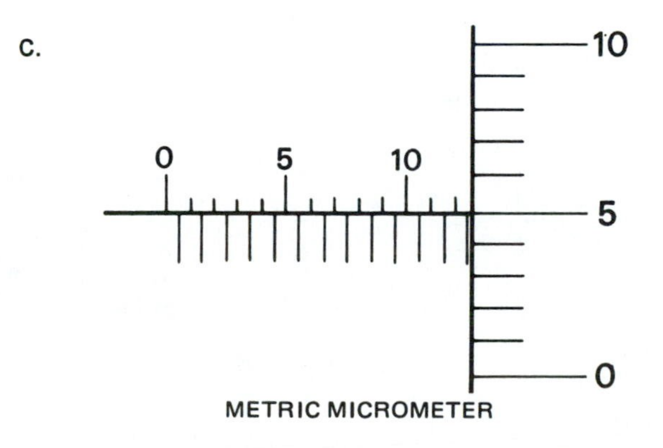

METRIC MICROMETER

c. ____________

d.

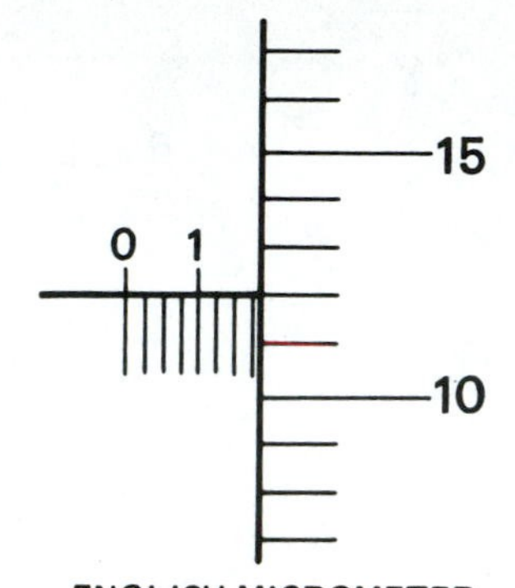
15
0 1
10
ENGLISH MICROMETER

d. ____________

e.

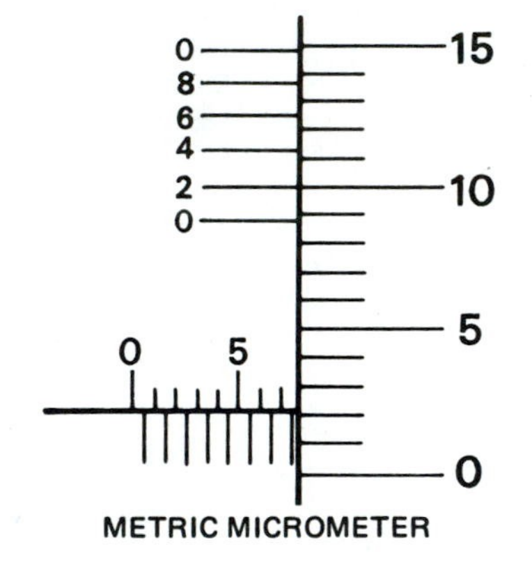
0
8
6
4
2
0
15
10
0 5
5
0
METRIC MICROMETER

e. ____________

f.

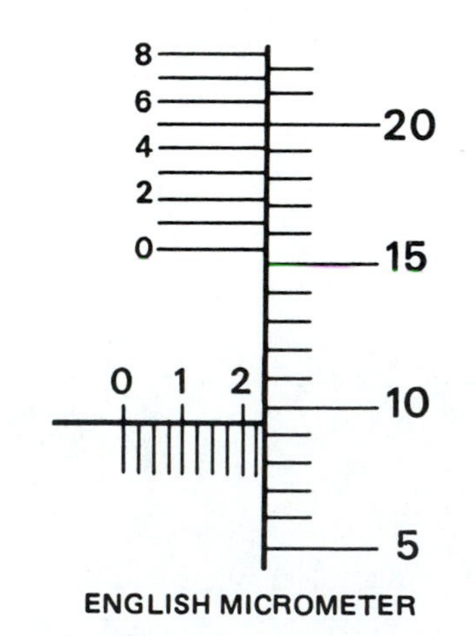
8
6
4
2
0
20
15
0 1 2
10
5
ENGLISH MICROMETER

f. ____________

g.

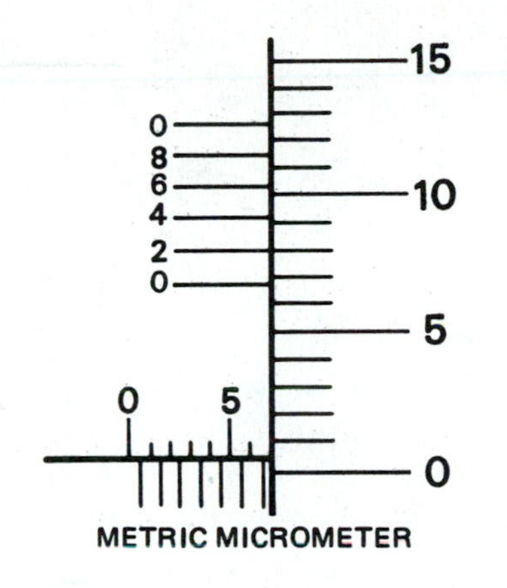
15
0
8
6
10
4
2
0
5
0 5
0
METRIC MICROMETER

g. ____________

h.

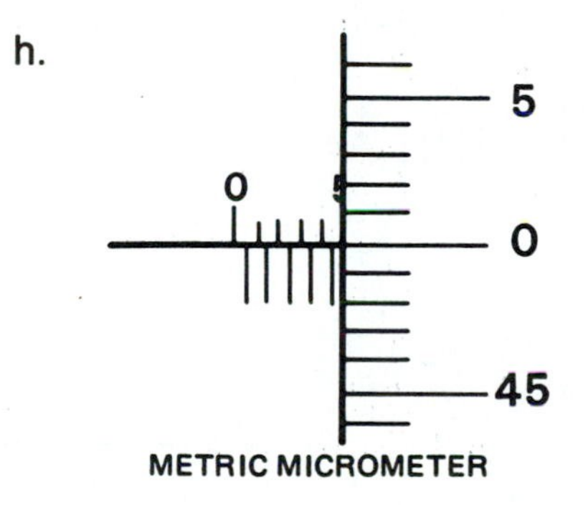
5
0 5
0
45
METRIC MICROMETER

h. ____________

i.

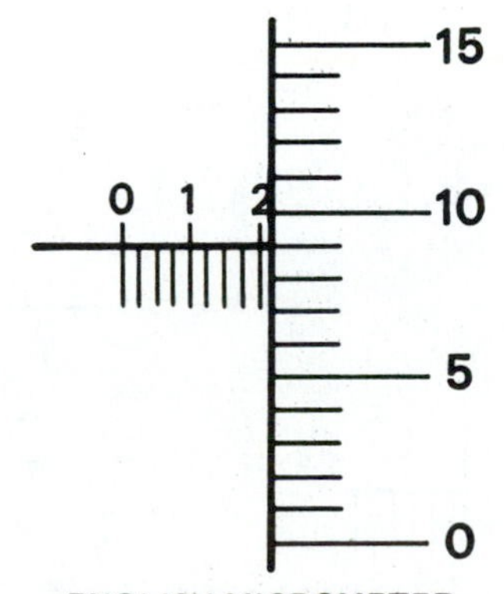
15
0 1 2
10
5
0
ENGLISH MICROMETER

i. ____________

j. 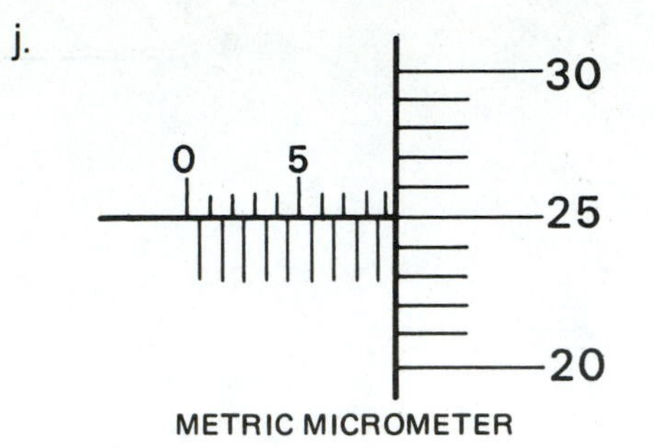

j. ____________

4. Determine the values of each vernier caliper reading.

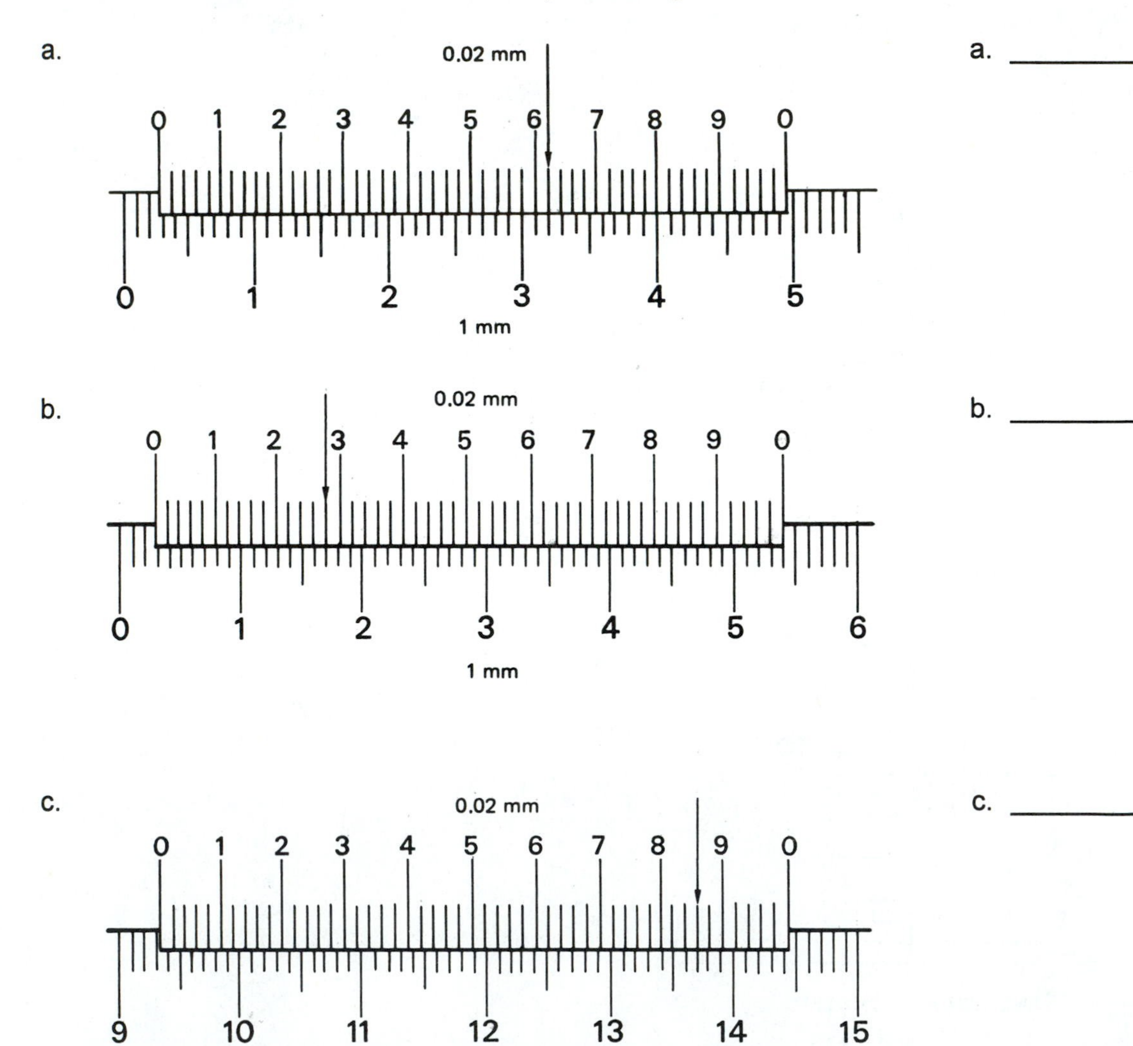

d.

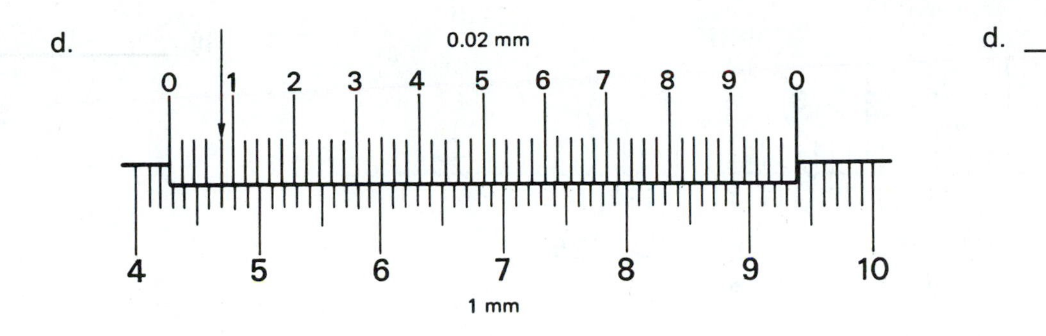
0.02 mm
0 1 2 3 4 5 6 7 8 9 0
4 5 6 7 8 9 10
1 mm

d. ____________

e.

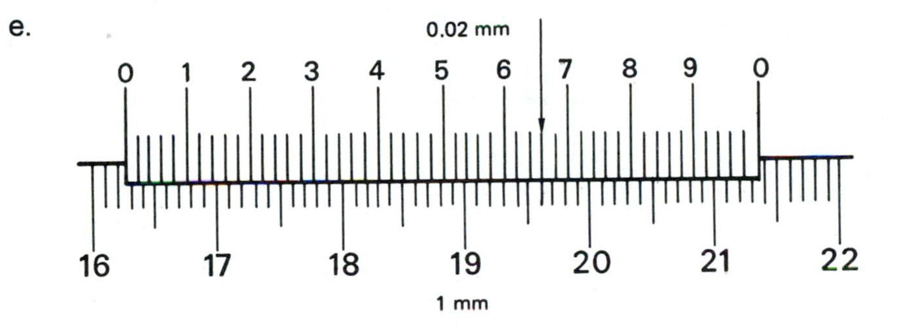
0.02 mm
0 1 2 3 4 5 6 7 8 9 0
16 17 18 19 20 21 22
1 mm

e. ____________

f.

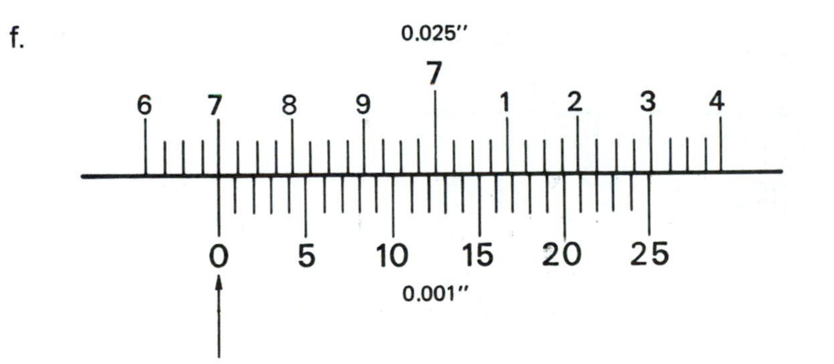
0.025″
7
6 7 8 9 1 2 3 4
0 5 10 15 20 25
0.001″

f. ____________

g.

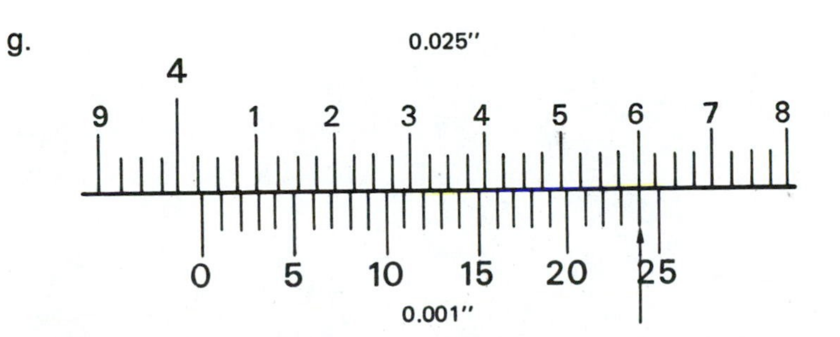
0.025″
4
9 1 2 3 4 5 6 7 8
0 5 10 15 20 25
0.001″

g. ____________

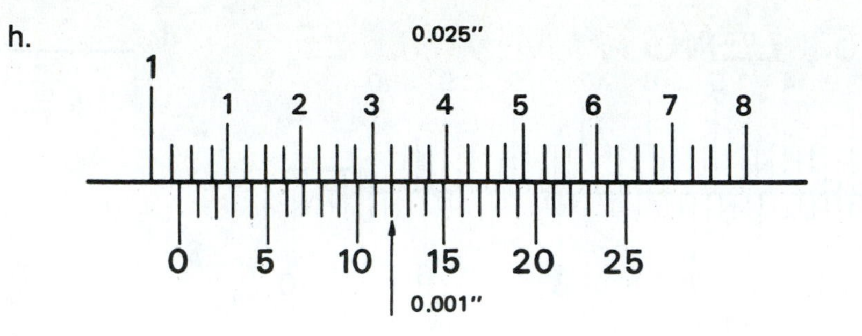
h.
0.025"
1
1 2 3 4 5 6 7 8
0 5 10 15 20 25
0.001"

h. ____________

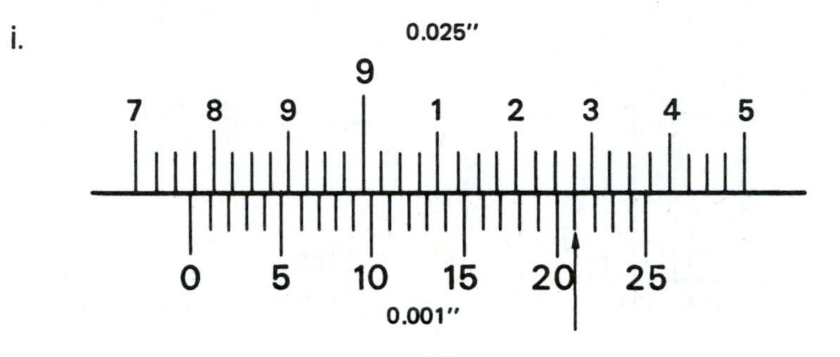
i.
0.025"
9
7 8 9 1 2 3 4 5
0 5 10 15 20 25
0.001"

i. ____________

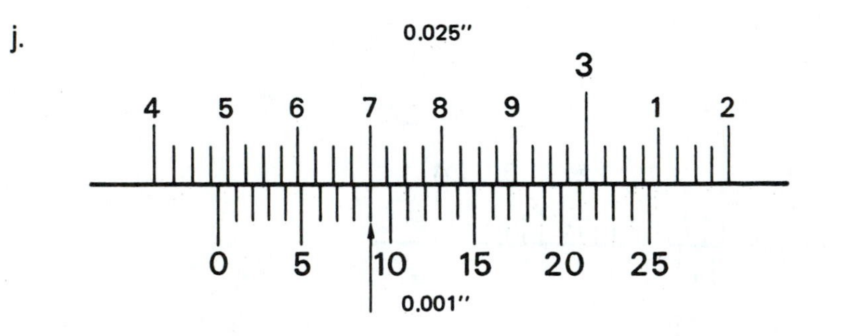
j.
0.025"
3
4 5 6 7 8 9 1 2
0 5 10 15 20 25
0.001"

j. ____________

Unit 16 LENGTH MEASURE

BASIC PRINCIPLES

COMMON ENGLISH LINEAR UNITS

12 inches	=	1 foot
3 feet	=	1 yard
$16\frac{1}{2}$ feet	=	1 rod
5,280 feet	=	1 mile
1,760 yards	=	1 mile

COMMON METRIC LINEAR UNITS

1 millimeter	=	0.001 meter
1 centimeter	=	0.01 meter
1 decimeter	=	0.1 meter
1 meter	=	1.0 meter
1 dekameter	=	10.0 meters
1 hectometer	=	100.0 meters
1 kilometer	=	1,000.0 meters

COMMON CONVERSION FACTORS

1 inch	=	25.4 millimeters
1 foot	=	0.3048 meter
1 yard	=	0.9144 meter

Length is probably the dimension most often measured by machinists. There are two systems used today in the United States: 1) the English system, which has been the traditional system used, and 2) the metric system, which is becoming more and more popular in industry. The English system is based on the inch, whereas the metric system is based on the meter. The English system uses fractional parts of an inch, such as $\frac{1}{2}$, $\frac{1}{4}$, $\frac{1}{8}$, $\frac{1}{16}$, $\frac{1}{32}$, and $\frac{1}{64}$. The metric system uses parts of a meter. For example, 1 decimeter is $\frac{1}{10}$ of a meter, and 1 kilometer is 1,000 meters. Since it is based on tens, the metric system tends to be easier to use, especially when conversion within the system is involved.

Example: Convert 3 feet to inches.

1 foot = 12 inches, therefore 3 x 12 inches = 36 inches

Example: Convert 10 inches to millimeters.

1 inch = 25.4 mm, therefore 10 x 25.4 mm = 254 mm

PRACTICAL PROBLEMS

1. Express 3 feet as inches. ______
2. Express 2 feet 6 inches as inches. ______
3. Express 1 foot $5\frac{3}{4}$ inches as inches. ______
4. Express 3 feet $4\frac{7}{16}$ inches as inches. ______
5. Express 11.9 feet as feet and inches. Express the inches to the nearer 64th inch. ______
6. Express 12.71 feet as feet and inches. Express the inches to the nearer 64th inch. ______
7. Express 2.65 feet as feet and inches. Express the inches to the nearer 64th inch. ______
8. A length of bar, 3 feet $7\frac{15}{16}$ inches long, is cut from a bar 12 feet $2\frac{3}{32}$ inches long. There is $\frac{1}{16}$ inch waste for the cut. What length of bar remains? Express answer in feet and inches. ______
9. A 20-foot bar of $\frac{3}{4}$-inch hexagonal steel is cut into three equal parts. How long is each part? Allow $\frac{1}{16}$ inch waste for each cut. Express answer in feet and inches. ______
10. How many arbors, each 1 foot $\frac{9}{16}$ inches long, can be cut from a piece of $1\frac{1}{4}$-inch diameter stock that is 12 feet 6 inches long? Allow $\frac{1}{16}$ inch waste for each cut. ______
11. How many millimeters are there in one decimeter? ______
12. How many millimeters are there in one meter? ______

13. How many centimeters are there in one meter? __________

14. How many millimeters are there in 7 decimeters? __________

15. How many millimeters are there in 85 centimeters? __________

Note: Use this diagram for problems 16–18.

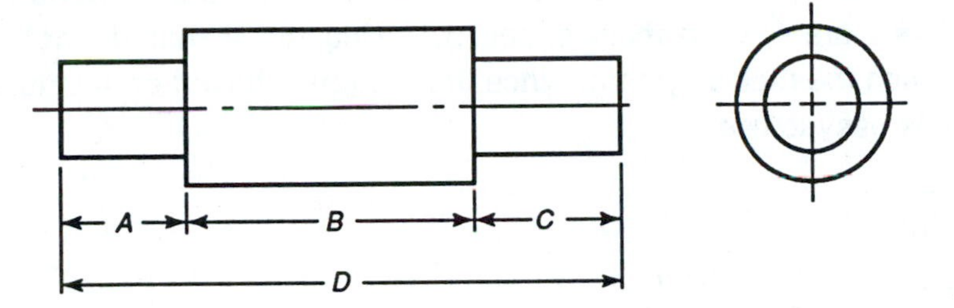

16. ***A*** = 35 millimeters; ***B*** = 110 millimeters; ***C*** = 45 millimeters. Find ***D***. __________

17. ***D*** = 45 centimeters; ***A*** = 75 millimeters; ***B*** = 150 millimeters. Find ***C***. __________

18. ***D*** = 3 decimeters; ***A*** = 81 millimeters; ***C*** = 93 millimeters. Find ***B***. __________

Note: Use this diagram for problems 19–22.

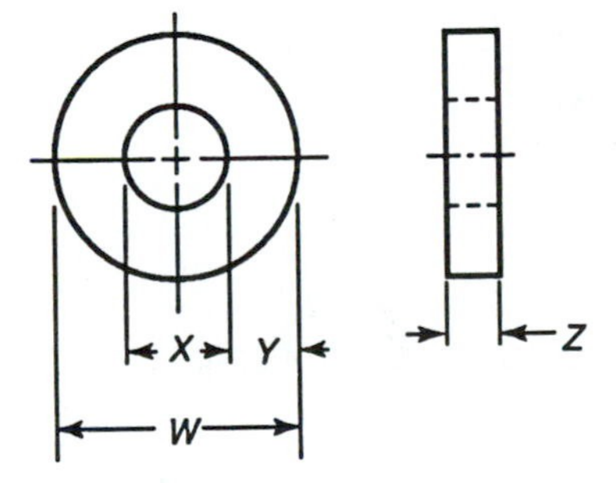

19. ***W*** = 95 millimeters; ***X*** = 45 millimeters. Find ***Y***. __________

20. ***W*** = 9 centimeters; ***X*** = 38 millimeters. Find ***Y***. __________

21. ***W*** = 2 decimeters; ***Y*** = 35 millimeters. Find ***X***. __________

22. ***X*** = 6 centimeters; ***Y*** = 22 millimeters. Find ***W***. __________

23. How many pieces of 5.0-millimeter drill rod, each 37 millimeters long, can be cut from a 1-meter-long bar? Allow 1.5 millimeters waste for each cut. __________

Unit 17 EQUIVALENT UNITS OF LENGTH MEASURE

BASIC PRINCIPLES

- During the last few years, the metric system has taken on greater importance in the United States. There are few machine shops operating today that do not have the capability to work with the metric system. Since the system is based on multiples of ten, the metric system is easy to use.

1 millimeter (mm)	=	0.03937 inch
1 centimeter (cm)	=	0.3937 inch
1 decimeter (dm)	=	3.937 inches
1 meter (m)	=	39.370 inches

1 inch	=	25.4 mm
1 inch	=	2.54 cm
1 inch	=	0.254 dm
1 inch	=	0.0254 m

PRACTICAL PROBLEMS

Note: For problems 1–3, round the answer to two decimal places, when necessary.

1. Express 5 inches as millimeters. __________
2. Express 1.562 inches as millimeters. __________
3. Express 3.750 inches as millimeters. __________

Note: For problems 4–11 round the answer to three decimal places, when necessary.

4. Express 25 millimeters as inches. __________
5. Express 19.72 millimeters as inches. __________
6. Express 312.451 millimeters as inches. __________

7. Express 4.5 centimeters as inches. ______

8. Express 0.109 decimeters as inches. ______

9. Express 2.17 centimeters as inches. ______

10. Express 118.110 inches as meters. ______

11. Express 19.981 inches as decimeters. ______

12. Eight pieces of stock, each 174 millimeters long, must be cut from a 10-foot bar of stock. Allowing 4 mm for waste, how many feet and inches of bar stock are left? Express the inches to the nearer 16th inch. ______

13. These lengths must be cut: 1 m, 0.27 m, 19 cm, 373 mm, and 4 dm. How long a bar must be used? Express the answer to the nearer foot and whole inch. ______

14. A square bar, 25 mm thick and 190 cm on each side, is cut into four equal pieces. How many inches wide is each piece? Round the answer to the nearer thousandth inch. ______

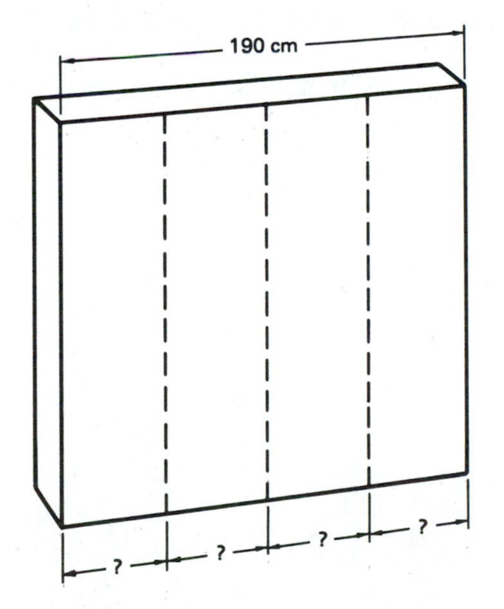

15. A machinist must make 19 shafts, each $11\frac{7}{8}$ inches long. How many decimeters of material will be needed? Round the answer to the nearer thousandth. ______

16. What is the diameter, in millimeters, of a pulley 18 inches in diameter? ______

17. What is the millimeter equivalent of 3 feet $3\frac{3}{4}$ inches? __________

18. How many 1-meter-long pieces and be cut from a 20-foot steel bar? __________

19. How many reamer blanks, each 20 centimeters long, can be cut from a bar 6 feet long? Allow 3 millimeters waste for each saw cut. __________

20. A ten-foot bar of $\frac{5}{8}$-inch diameter mild steel is divided into eight equal parts. Give the length of each part in millimeters. __________

21. Number *16* gauge sheet steel is $\frac{1}{16}$ inch thick. Find the thickness in millimeters. __________

22. A steel bar is 5 feet $11\frac{1}{2}$ inches long. What is its length in centimeters? __________

23. If seven pieces of stock, each 75 millimeters long, are cut from an 18-foot bar, what length is left, in feet and inches? Allow 3 millimeters of waste for each cut. Express the inches to the nearer 64th inch. __________

24. One job order calls for three pieces of stock, each 375 millimeters long. Another job order calls for two pieces of stock, each $5\frac{7}{8}$ inches long. Find, in millimeters, the difference between the total lengths of stock. __________

25. Find, in inches, the difference between 5 decimeters and 18 inches. __________

Unit 18 ANGULAR MEASUREMENT

BASIC PRINCIPLES

- In the English and metric systems angles are measured in degrees, minutes, and seconds.

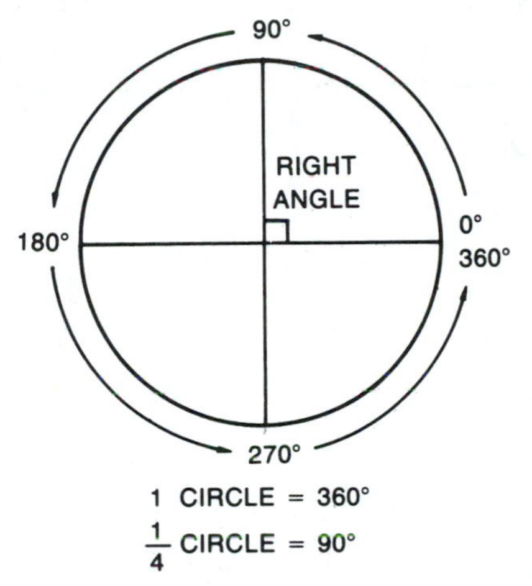

1 CIRCLE = 360°
$\frac{1}{4}$ CIRCLE = 90°

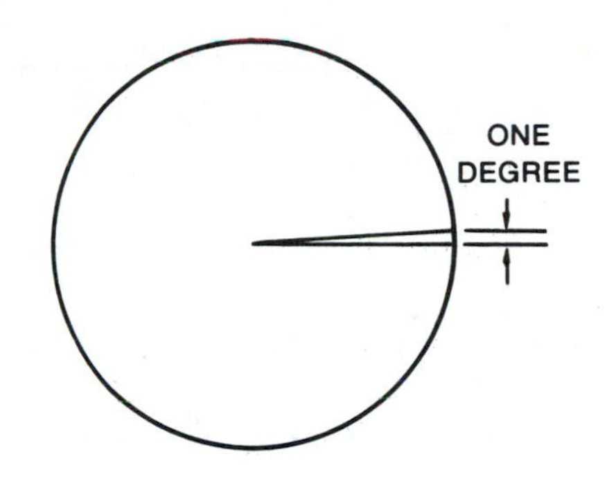

1 DEGREE (1°) = 60 MINUTES (60′)
1 MINUTE (1′) = 60 SECONDS (60″)

PRACTICAL PROBLEMS

1. How many degrees are there in a right angle? __________

2. How many degrees are there in two right angles? __________

3. How many 60-degree angles are there in two right angles? __________

Note: Use this diagram for problems 4-6.

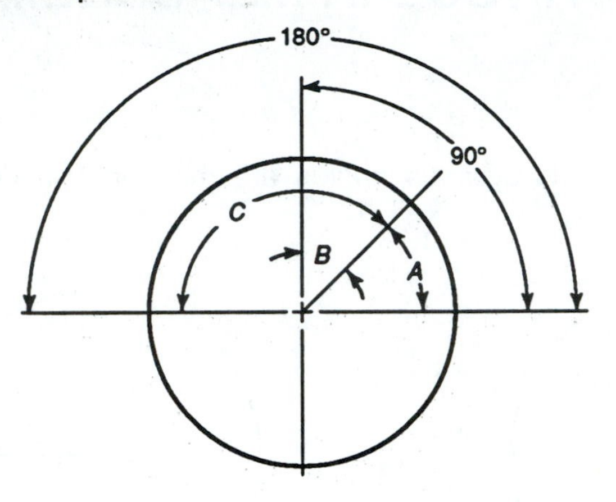

4. Angle ***A*** = 30 degrees. What is the value of angle ***B***? __________

5. Angle ***A*** = 30 degrees. Find the value of angle ***C***. __________

6. Angle ***A*** = 61 degrees 40 minutes. What is the value of angle ***B***? __________

7. Find the complement of 22 degrees 47 minutes. __________

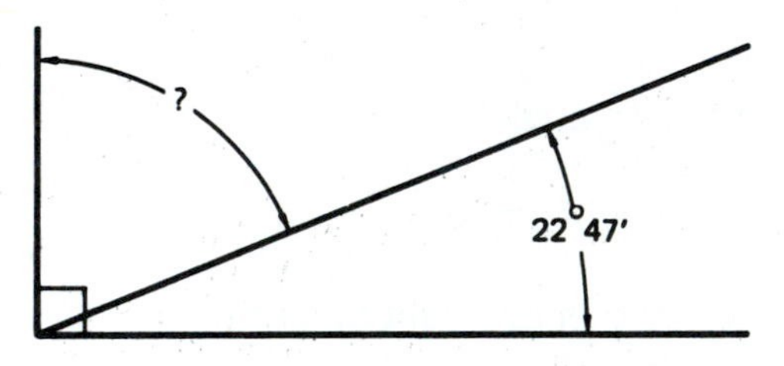

8. Find the supplement of 105 degrees 22 minutes. __________

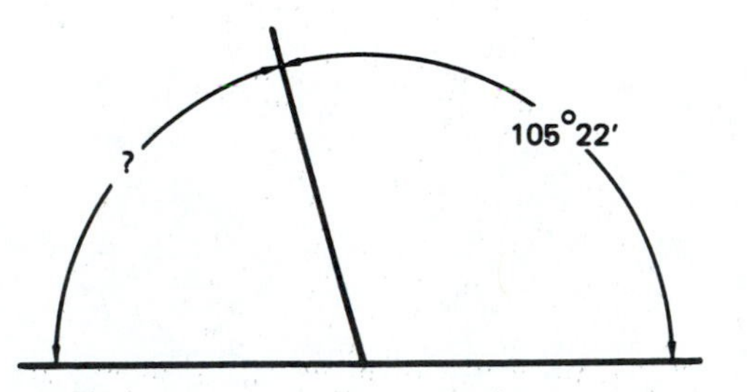

9. Find the angle whose complement is 22 degrees 30 minutes. __________

10. In this circle:

Angle ***A*** = 30 degrees

Angle ***B*** = 20 degrees

Angle ***D*** = 65 degrees 22 minutes

Angle ***F*** = 118 degrees 40 minutes

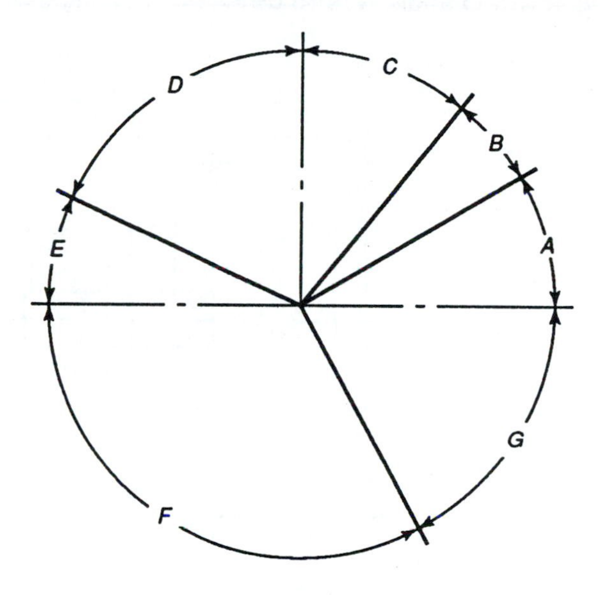

a. What is the sum of angles ***A*** and ***B***? a. ____________

b. Angles ***A***, ***B***, and ***C*** total 90 degrees. Find the value of angle ***C***. b. ____________

c. What is the sum of angles ***A***, ***B***, ***C***, and ***D***? c. ____________

d. Find the value of angle ***E***. d. ____________

e. What is the sum of angles ***A***, ***B***, ***C***, ***D***, ***E***, and ***F***? e. ____________

f. Find angle ***G***. f. ____________

11. The included angle ***R*** is 48 degrees. Find angle ***S***. ____________

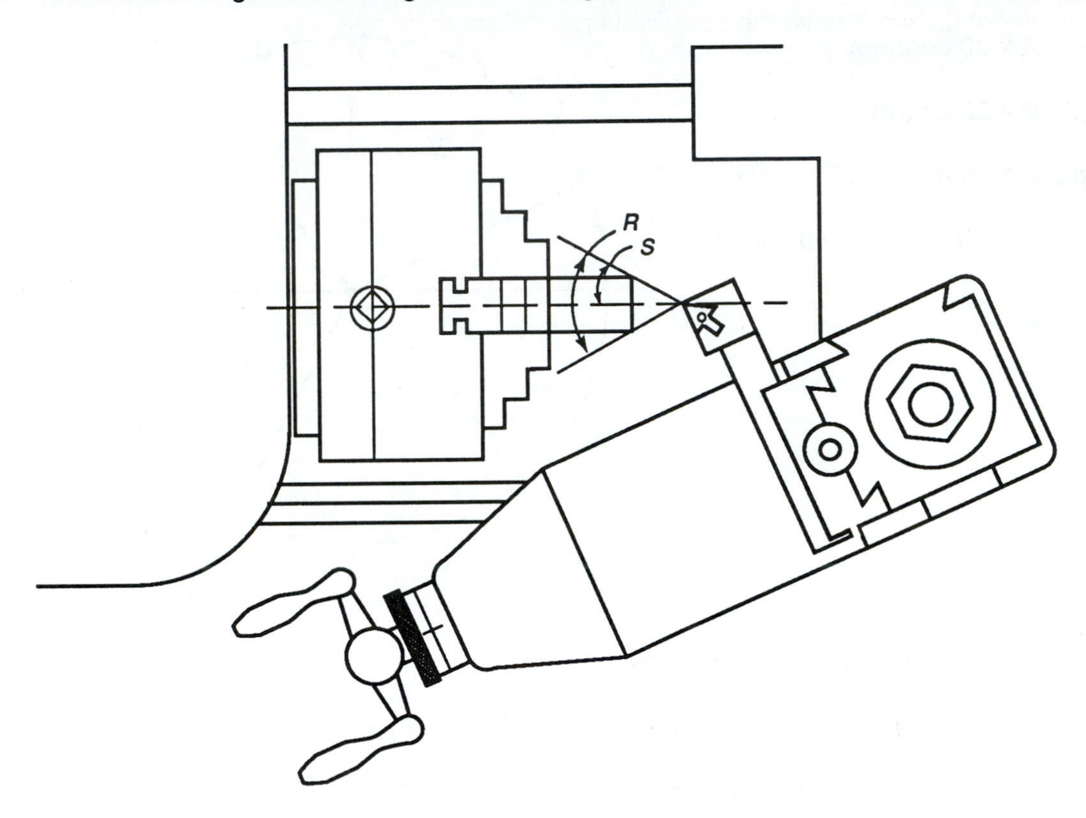

12. To cut an included angle of 72 degrees 30 minutes. how many degrees is the compound rest on a lathe moved? ____________

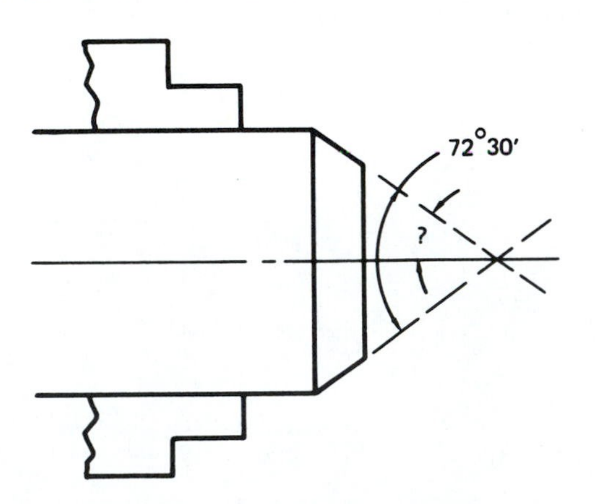

13. This piece of work is to be turned on a lathe. At what angle to the centerline of the work must the compound rest be set to turn an included angle of 120 degrees? ________

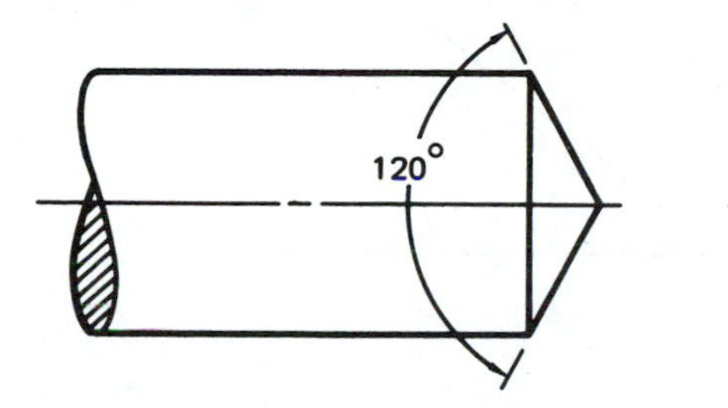

14. When turning a lathe center point, at what angle to the centerline of the lathe bed should the compound rest be set? ________

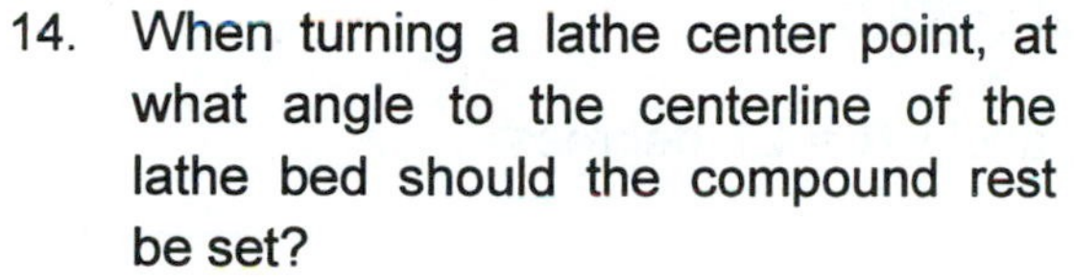

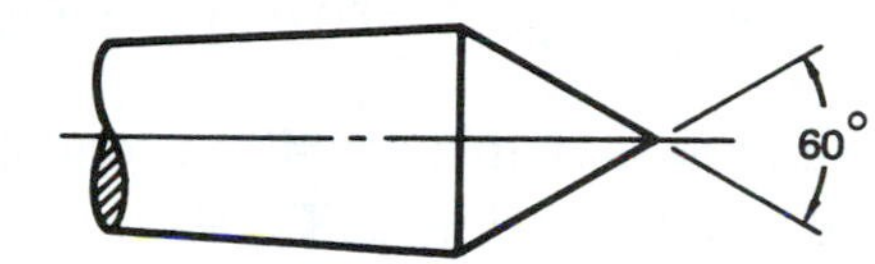

15. If the hexagonal stock is planed in a vise, at what angle from vertical is the head set? ________

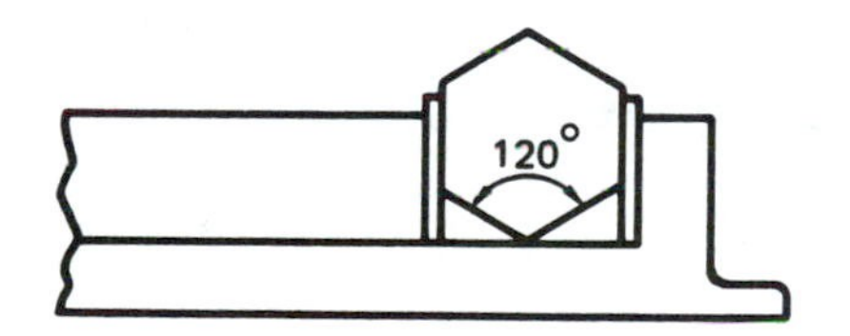

16. A layout of five holes equally spaced on a 10-inch circle is shown. What angle is formed by connecting the centers of holes ***B*** and ***C*** with the center? ________

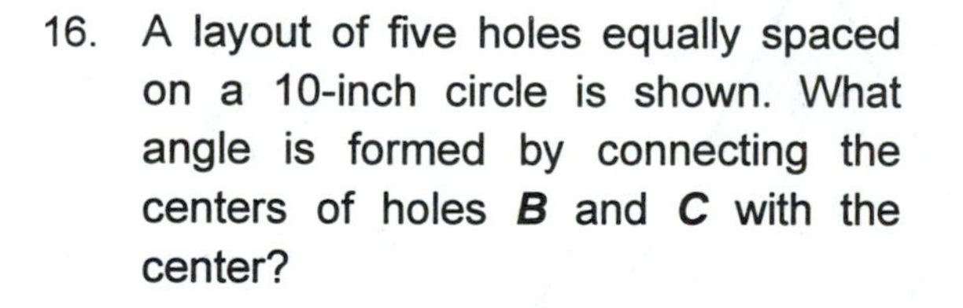

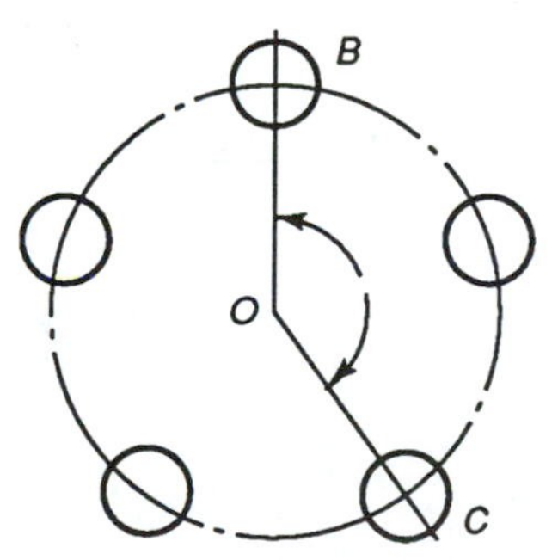

Note: Use this diagram for problems 17 and 18.

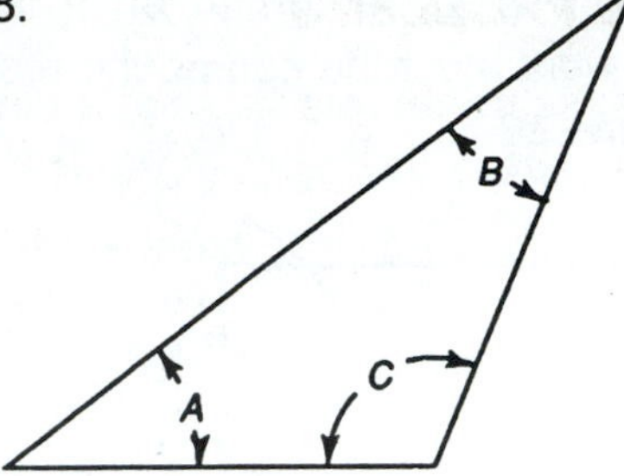

17. Angle ***A*** = 20 degrees; angle ***B*** = 40 degrees. Find angle ***C***. __________

18. Angle ***A*** = 17 degrees 21 minutes 43 seconds; angle ***B*** = 27 degrees 52 minutes 51 seconds. Find angle ***C***. __________

Note: Use this diagram for problems 19 and 20.

19. Angle ***D*** = 35 degrees.

    a. Find angle ***F***. a. __________

    b. Find angle ***E***. b. __________

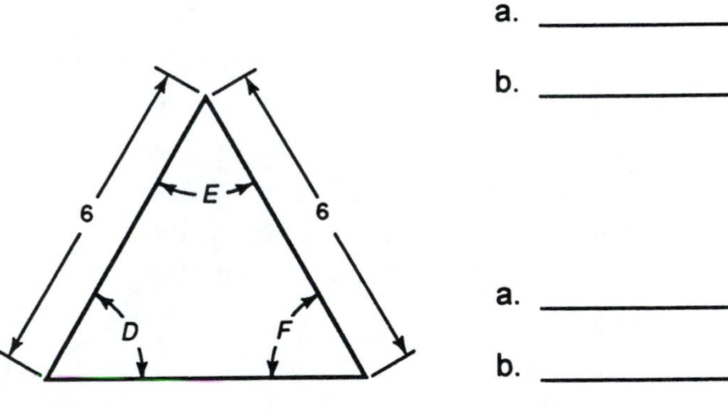

20. Angle ***F*** = 19 degrees 41 minutes.

    a. Find angle ***D***. a. __________

    b. Find angle ***E***. b. __________

21. A pulley has twelve spokes. What is the angle between the centerlines of any two adjacent spokes? __________

CRITICAL THINKING PROBLEMS

1. Using a scale, measure the diameter of a dime, a nickel, and a quarter. Record your answers to the nearest 1/64 inch. Compare your answers with other students' answers. How do they compare?

Dime ________

Nickel ________

Quarter ________

2. A regular pentagon has five equal sides and angles. Determine angle ***A***.

A = ________

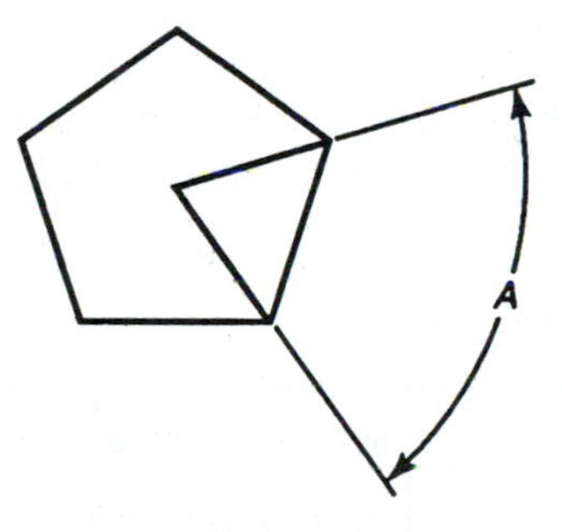

3. Using any method, determine angle ***A*** and dimension ***C*** on a regular hexagon.

A = ________

C = ________

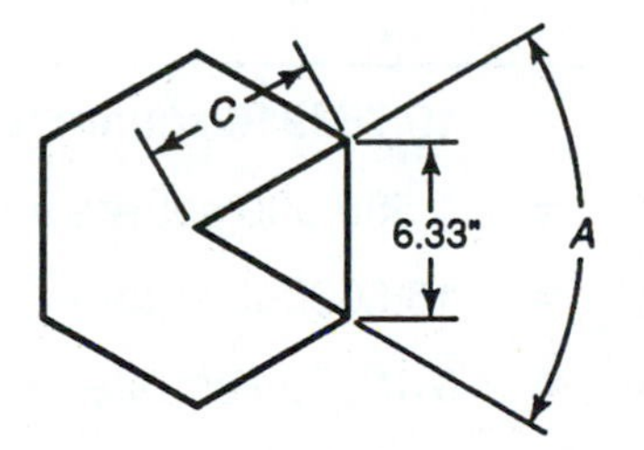

Computed Measure

Unit 19 SQUARE MEASURE

BASIC PRINCIPLES

- Area is always computed in square units. When computing area all lengths must be in the same units.

1 square yard (sq. yd.)	=	9 square feet (sq. ft.)
1 square foot (sq. ft.)	=	144 square inches (sq. in.)

100 square millimeters (mm^2)	=	1 square centimeter (cm^2)
100 square centimeters (cm^2)	=	1 square decimeter (dm^2)
100 square decimeters (dm^2)	=	1 square meter (m^2)

1 square meter (m^2)	=	10.763910 square feet (sq. ft.)
1 square meter (m^2)	=	1,550.000 square inches (sq. in.)
1 square decimeter (dm^2)	=	15.500000 square inches (sq. in.)
1 square centimeter (cm^2)	=	0.155000 square inch (sq. in.)
1 square millimeter (mm^2)	=	0.001550 square inch (sq. in.)

1 square foot (sq. ft.)	=	0.092 903 square meter (m$^{2)}$
1 square inch (sq. in.)	=	0.000 645 square meter (m^2)
1 square inch (sq. in.)	=	0.064 516 square decimeters (dm^2)
1 square inch (sq. in.)	=	6.451 600 square centimeters (cm^2)
1 square inch (sq. in.)	=	645.160 square millimeters (mm^2)

PRACTICAL PROBLEMS

1. Express 8 sq. ft. as square inches. __________
2. Express 2,592 sq. in. as square yards. __________
3. Express 936 sq. in. as square feet. __________
4. Express 30 m^2 as square decimeters. __________
5. Express 1,650 mm^2 as square decimeters. __________
6. Express 950,000 mm^2 as square centimeters. __________

Note: For problems 7–10, round the answer to the nearer thousandth, when necessary.

7. Express 0.87 m^2 as square inches. __________
8. Express 1,950 mm^2 as square inches. __________
9. Express 0.83 m^2 as square feet. __________
10. Express 19 dm^2 as square inches. __________

Note: For problems 11–14, round the answer to four decimal places when necessary.

11. Express 11 sq. in. as square meters. __________
12. Express 0.75 sq. ft. as square meters. __________
13. Express 12.3 sq. in. as square centimeters. __________
14. Express 15.73 sq. in. as square decimeters. __________

Unit 20 AREA OF SQUARES, RECTANGLES, AND PARALLELOGRAMS

BASIC PRINCIPLES

- Area is expressed in *square* units.
- The area of a square is equal to the length of one side times itself.

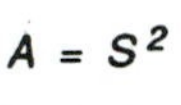

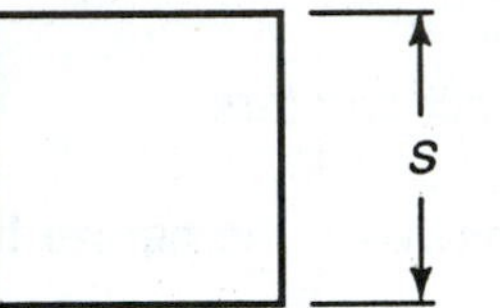

- The area of a rectangle is equal to the length times the height.

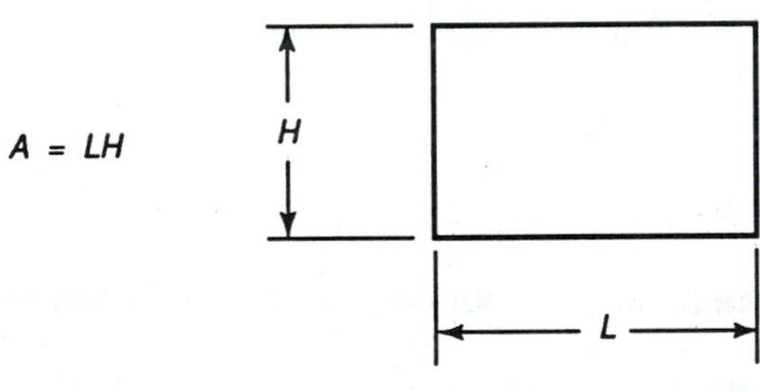

- The area of a parallelogram is equal to the base times the height.

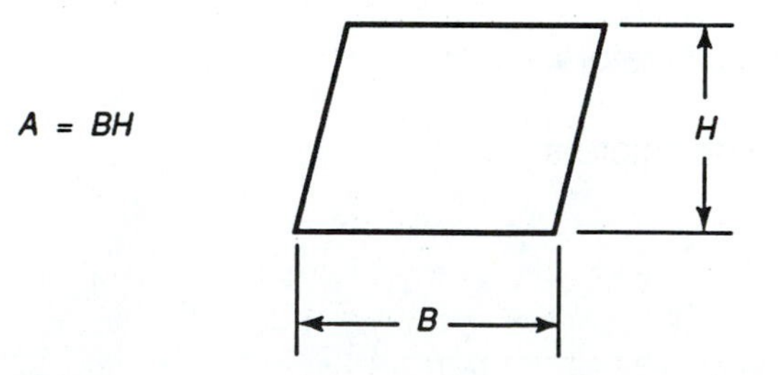

PRACTICAL PROBLEMS

1. The tool crib in a machine shop is 14 feet long and $9\frac{1}{2}$ feet wide. How many square feet of floor space does it occupy? __________

2. How many square centimeters are in this piece of sheet metal? ____________

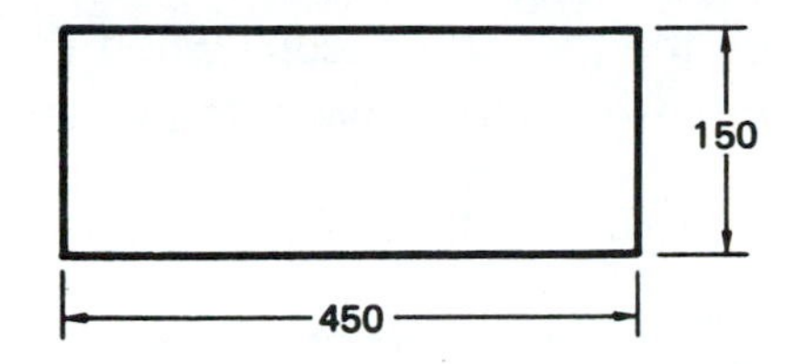

3. A person working on a surface grinder is paid 8 cents per square inch of finished surface. The pieces that are being ground are rectangular in shape and are $4\frac{1}{4}$ inches long by $2\frac{3}{4}$ inches wide. An average of $8\frac{1}{2}$ pieces are ground each hour. Find, to the nearer whole cent, the amount the worker earns in an 8-hour day. ____________

4. A piece of sheet metal in the shape of a parallelogram is 3.5 meters long and its altitude is 0.85 meter. Find the area in square meters. ____________

5. This rectangular piece of sheet metal has a parallelogram punched out of its center.

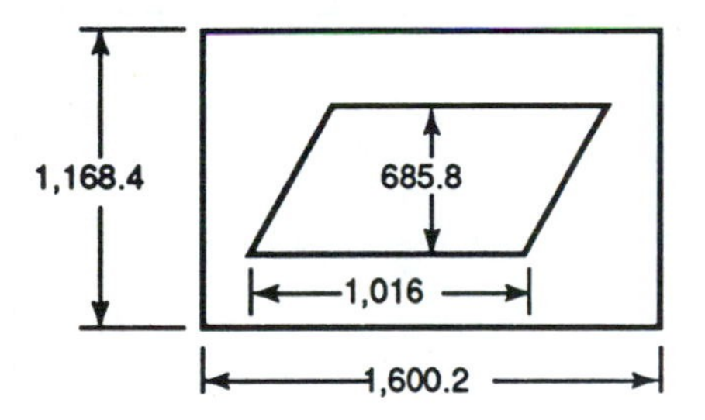

a. Find, in square millimeters, the area of the piece that is punched out. a. ____________

b. Find, in square millimeters, the area of the part that is left. b. ____________

6. Four sections of wire grill were set up to enclose an area for a tool crib. Later the wire grills are moved and set up as a parallelogram. How many more square feet of floor space does the tool crib occupy when it is a square than when it is a parallelogram? Round the answer to the nearer hundredth. ______

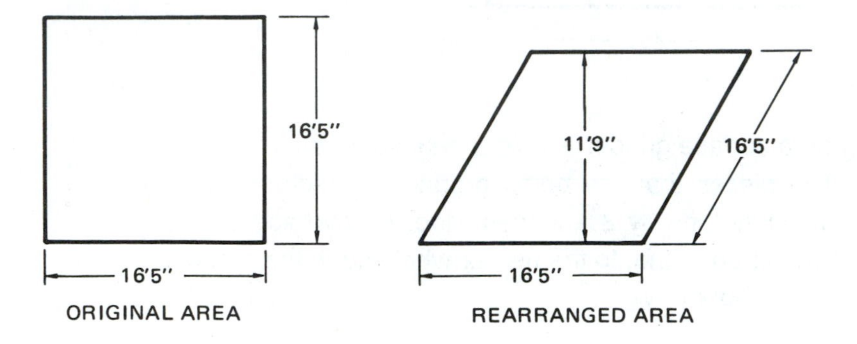

7. This 30-inch by 18-inch motor bracket must be made from $\frac{1}{2}$-inch plate. How many square feet of plate must be ordered for five brackets? ______

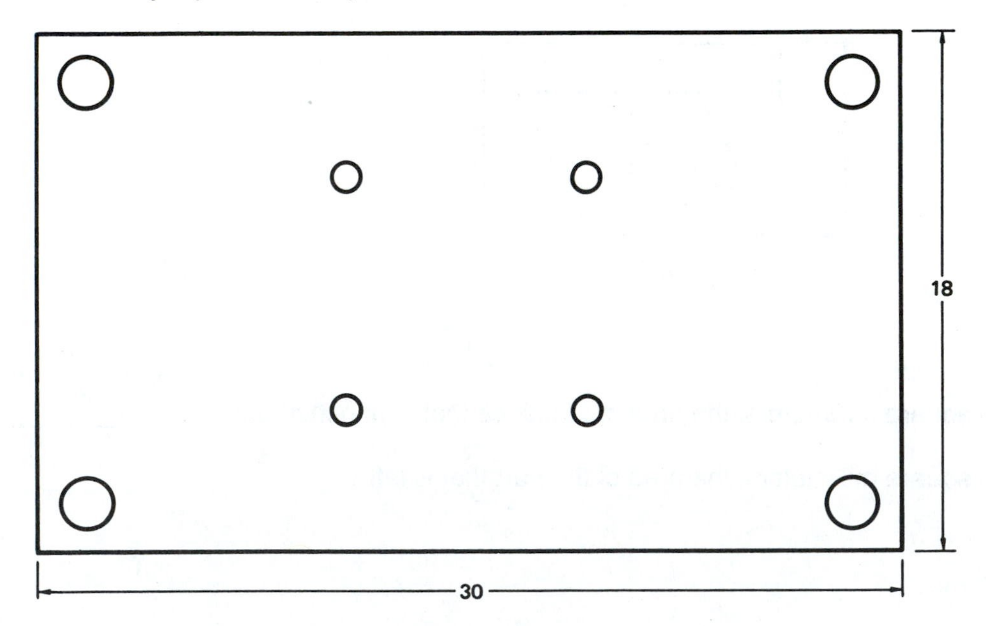

8. The dimensions of this part are in centimeters. How many square millimeters are removed from the cut away portion of this part? __________

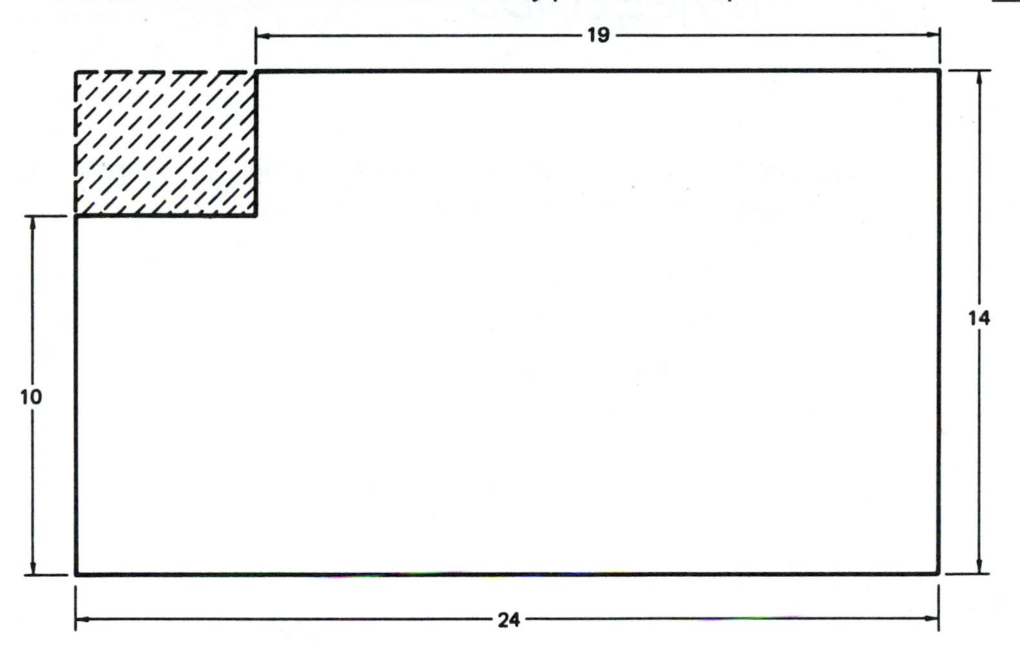

9. How many square inches are removed from the cut away portion of this part? __________

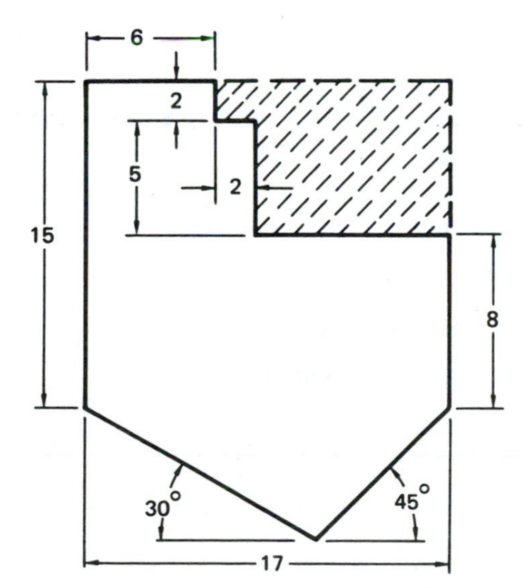

Unit 21 AREA OF TRIANGLES AND TRAPEZOIDS

BASIC PRINCIPLES

- If it is remembered that a triangle is simply one half of a rectangle or a parallelogram, it can be seen that the area of a triangle can be found by using the formula:

$$\text{Area} = \frac{1}{2}BH$$

where: area = square units
B = base
H = height

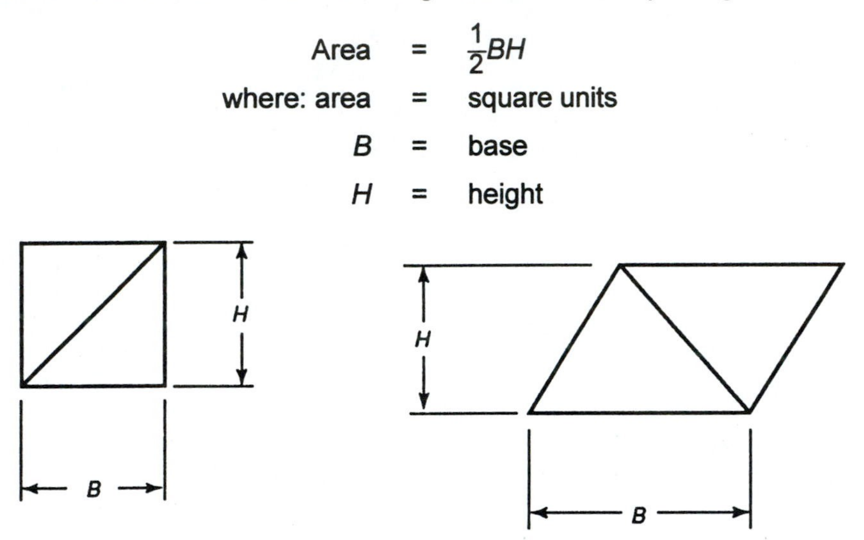

- A trapezoid is a four-sided figure with two parallel sides. The formula used to find the area of a trapezoid is as follows:

$$\text{Area} = \frac{H\,(B_1 + B_2)}{2}$$

where: area = square units
B = base
H = height

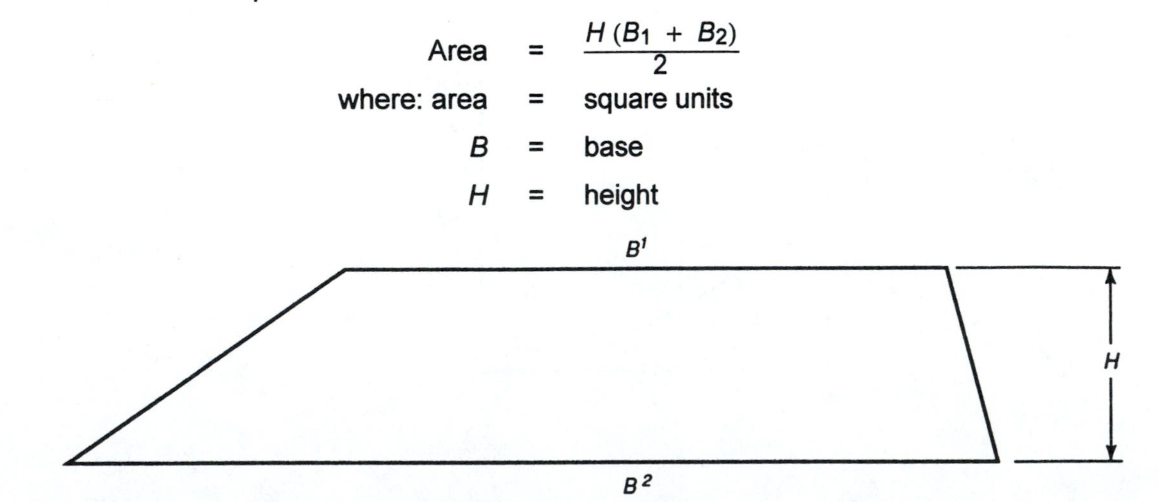

PRACTICAL PROBLEMS

1. A triangle with a base of 27 inches and an altitude of 17 inches is cut from a four-foot-square piece of sheet metal. Find, in square inches, the area of the sheet metal left in the piece. __________

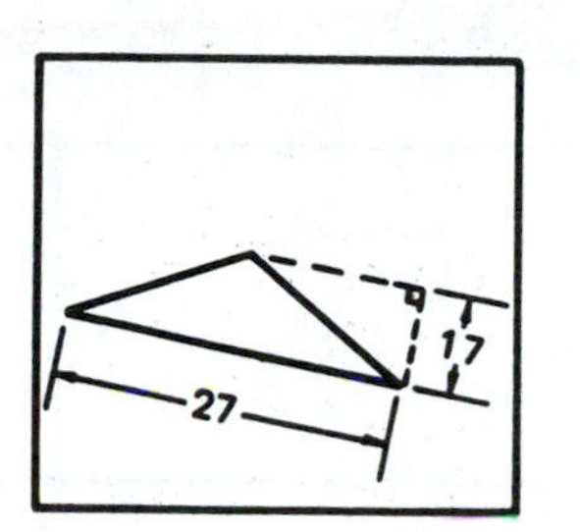

2. The base of a triangular piece of sheet metal is 1 foot. If the piece contains 100 square inches, what is the height? Round the answer to the nearer thousandth inch. __________

3. A triangular steel plate of the size and shape shown is to be carburized on one side. Find, to the nearer thousandth square inch, the area to be carburized. __________

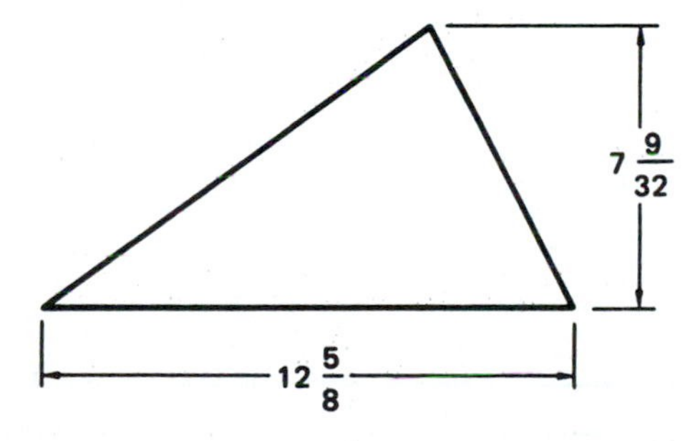

4. Triangular shaped pieces with a base of $6\frac{1}{16}$ inches and altitude of $8\frac{3}{8}$ inches are blanked out of a strip $8\frac{1}{2}$ inches wide. Eight pieces are blanked out of a strip 50 inches long.

   a. Find, to the nearer thousandth square inch, the area of the pieces. a. __________

   b. Find, to the nearer thousandth square inch, the area of the metal that is not used. b. __________

5. Find, to the nearer thousandth square inch, the area of this piece. ____________

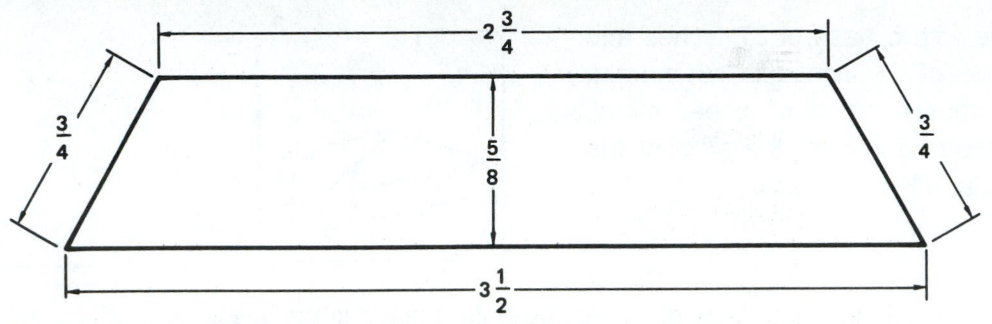

6. Find the areas of each trapezoid. Round the answer to the nearer thousandth when necessary.

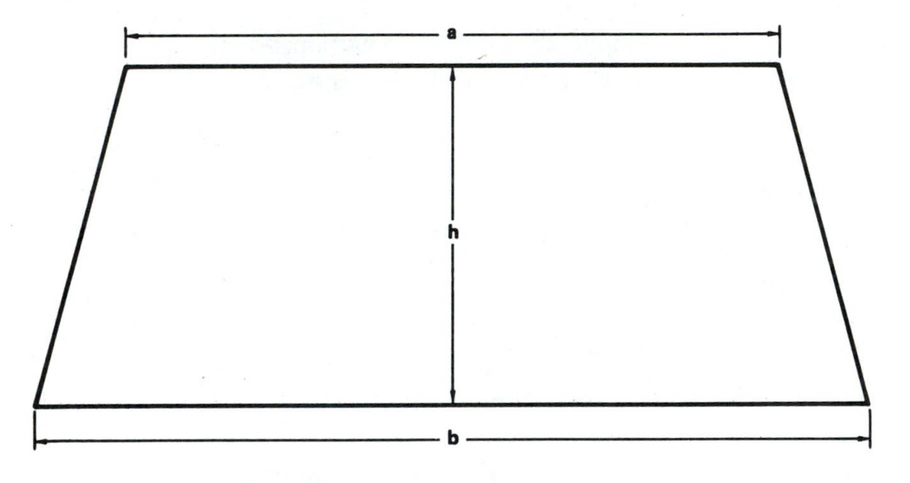

	h	a	b	AREA
a.	107 mm	210 mm	340 mm	
b.	2 1/2″	4 1/16″	6 5/16″	
c.	0.25 m	0.9 m	1.95 m	
d.	9.5″	13.75″	18.625″	

a. ____________

b. ____________

c. ____________

d. ____________

Unit 22 AREA OF CIRCULAR FORMS

BASIC PRINCIPLES

- Two common formulas for finding the area of a circle are

$$\text{Area} = \pi r^2 \quad \text{or} \quad \text{Area} = 0.7854D^2$$

where: area = square units

π = 3.1416

r = radius of the circle

D = diameter of the circle

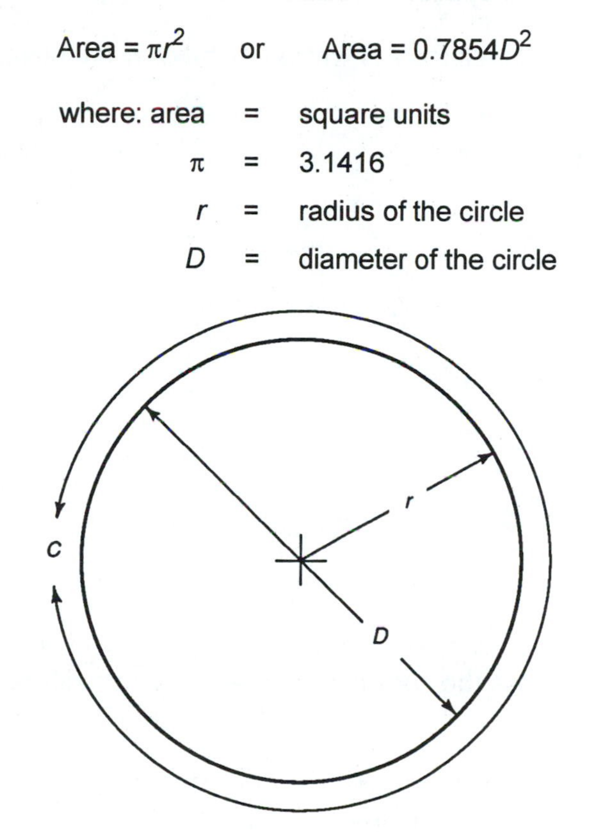

PRACTICAL PROBLEMS

1. A steel disk, 210 mm in diameter by 20 mm thick, is to be carburized on its flat surfaces only. Find the area that is to be carburized. ____________

2. A circular blower opening is covered by a heavy wire screen. The diameter of the opening is 1.52 meters. The screen is cut from a 2-meter-square piece.

   a. Find, to the nearer thousandth square meter, the area of the piece that is used to cover the opening. a. ________

   b. Find the area of the material that is wasted to the nearer thousandth square meter. b. ________

Use this diagram for problems 3–5.

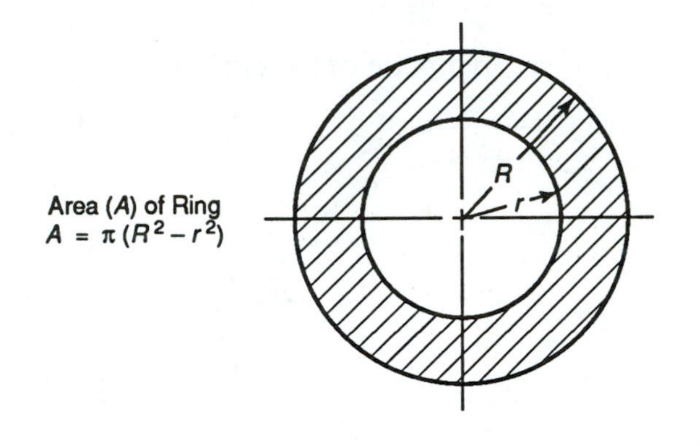

3. Find the area of a ring with an outside diameter of 110 mm and an inside diameter of 60 mm. Express the answer to the nearer hundredth square millimeter. ________

4. The outside diameter of a flywheel rim is $3\frac{1}{2}$ feet. The inside diameter of the rim is 2 feet 10 inches. Find the area of the rim to the nearer hundredth square inch. ________

5. A 14-inch diameter grinding wheel has a center hole 2 inches in diameter. Find, to the nearer thousandth square inch, the area of the wheel. ________

Unit 23 AREA OF CYLINDRICAL FORMS

BASIC PRINCIPLES

- The total outside surface area of an open cylinder (tube) may be calculated using the following formula:

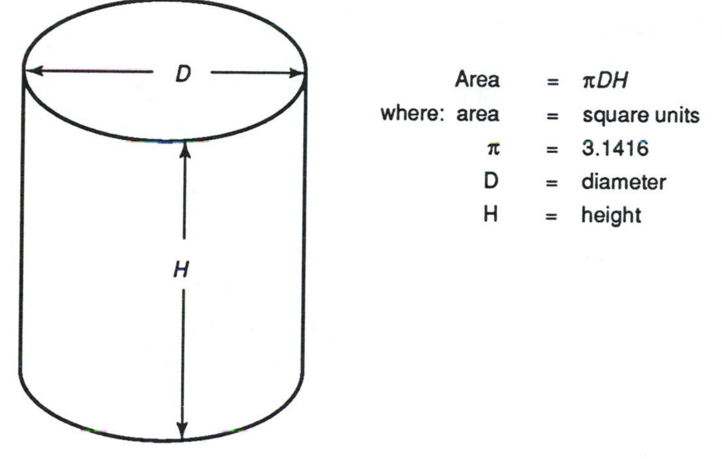

Area = πDH
where: area = square units
π = 3.1416
D = diameter
H = height

- If the sidewall of a cylinder is cut and unrolled, a rectangle is formed. When the circumference becomes the length of the rectangle, it is equivalent to pi (3.14) x diameter.

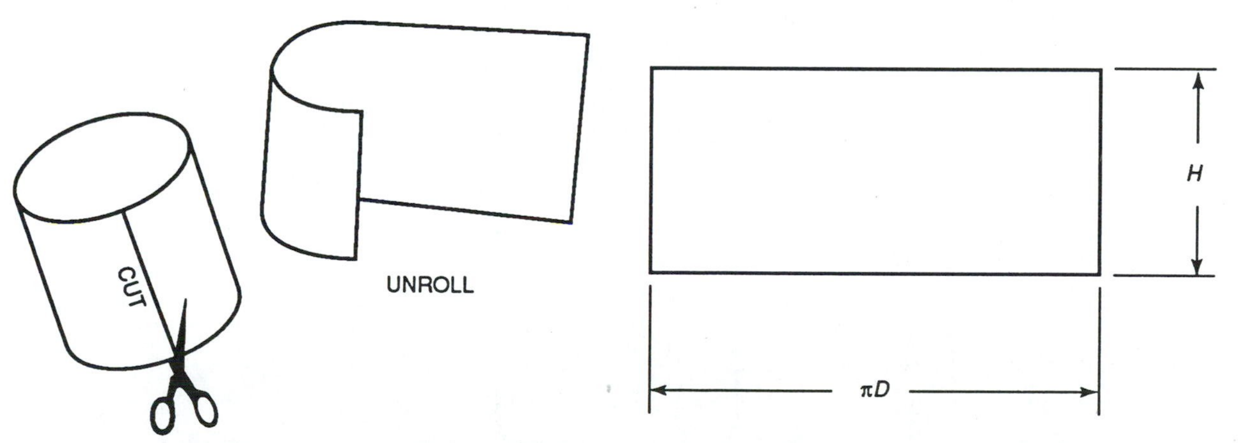

- If the cylinder is closed on the ends like a barrel, to find the total outside surface area one must calculate the area of the cylindrical portion and the area of the ends also. Therefore, the formula for the area of a circle is used to calculate the area of one end. That result is then doubled and added to the area of the cylindrical portion.

$$\text{Area of one end} = 0.7854D^2$$

Example: Find the surface area of a pipe 10 feet long x 2 feet in diameter.

$$\text{Surface area} = \pi DH = 3.1416(2)(10) = 62.832 \text{ ft}^2$$

If ends are to be put on this pipe to form a closed cylinder, the area of the ends is calculated as follows:

$$\text{Area of one end} = 0.7854D^2 = 0.7854(2^2) = 0.7854(4) = 3.1416 \text{ ft}^2$$

Since there are two ends, we must multiply the area of one end times two giving us 6.2832 ft^2 as the total area for the two ends. This figure is then added to the surface area of the open pipe to obtain a total surface area of 69.1152 ft^2 for a pipe with closed ends.

PRACTICAL PROBLEMS

1. What is the area of an end cap for this cylindrical tube? Express the answer to the nearer thousandth decimeter. ____________

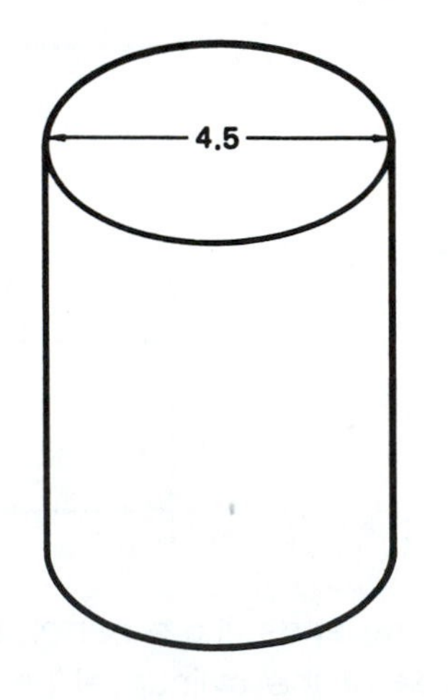

2. How many square meters of plate must be used to form this tube? Express the answer to the nearer thousandth. __________

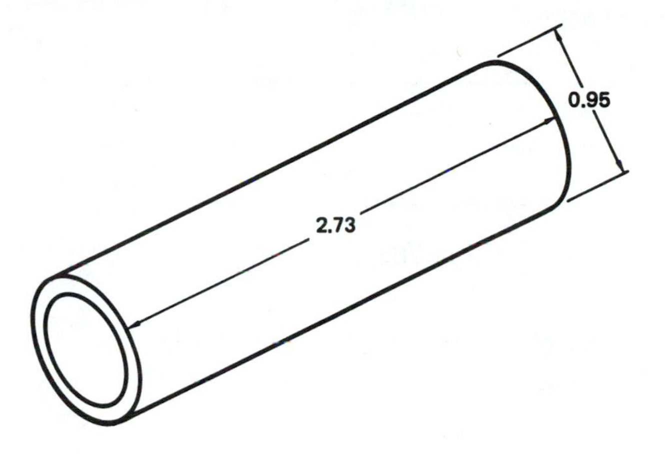

3. Find, to the nearer thousandth square centimeter, the amount of sheet metal that must be used for this container. __________

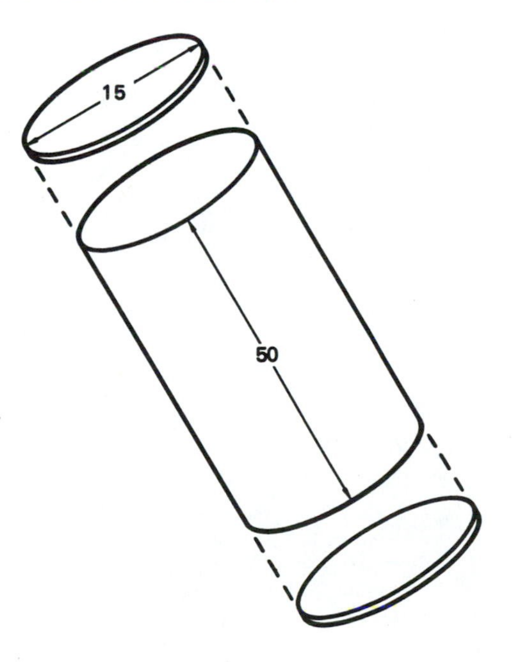

4. A shaft, 19 cm in diameter and 150 cm long, must be completely case hardened. What is the total area of the surface to be hardened to the nearer thousandth square centimeter? ____________

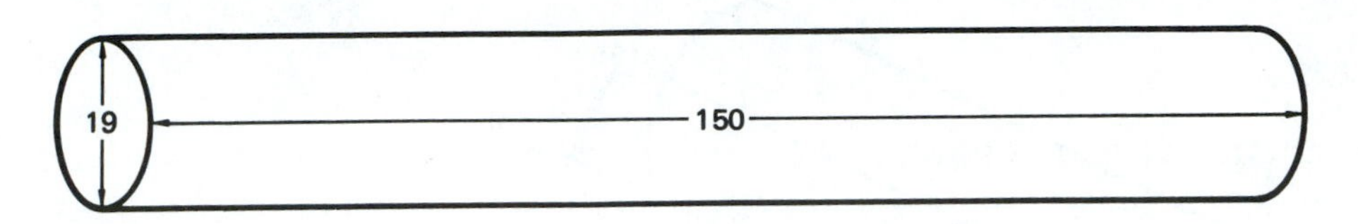

5. How many square inches of grinding wheel are in contact with the workpiece in each revolution of the wheel? ____________

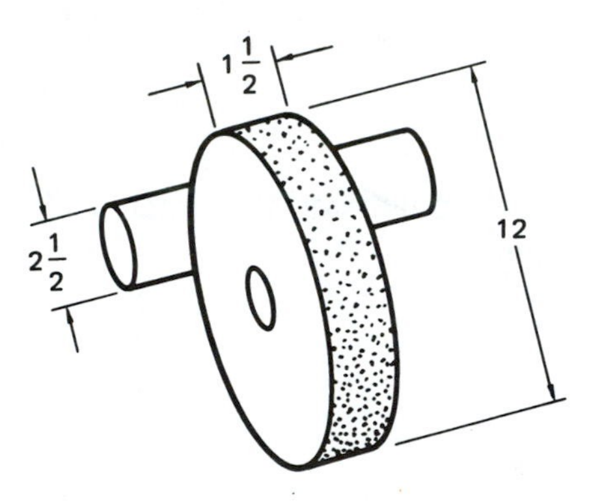

Unit 24 VOLUME OF RECTANGULAR SOLIDS

BASIC PRINCIPLES

- The volume of a rectangle is equal to the length times the height times the width and is expressed in *cubic* units. Area measurement uses two dimensions, whereas volume measurement uses three dimensions. As in area measure, all dimensions must be expressed in the same units.

Example:

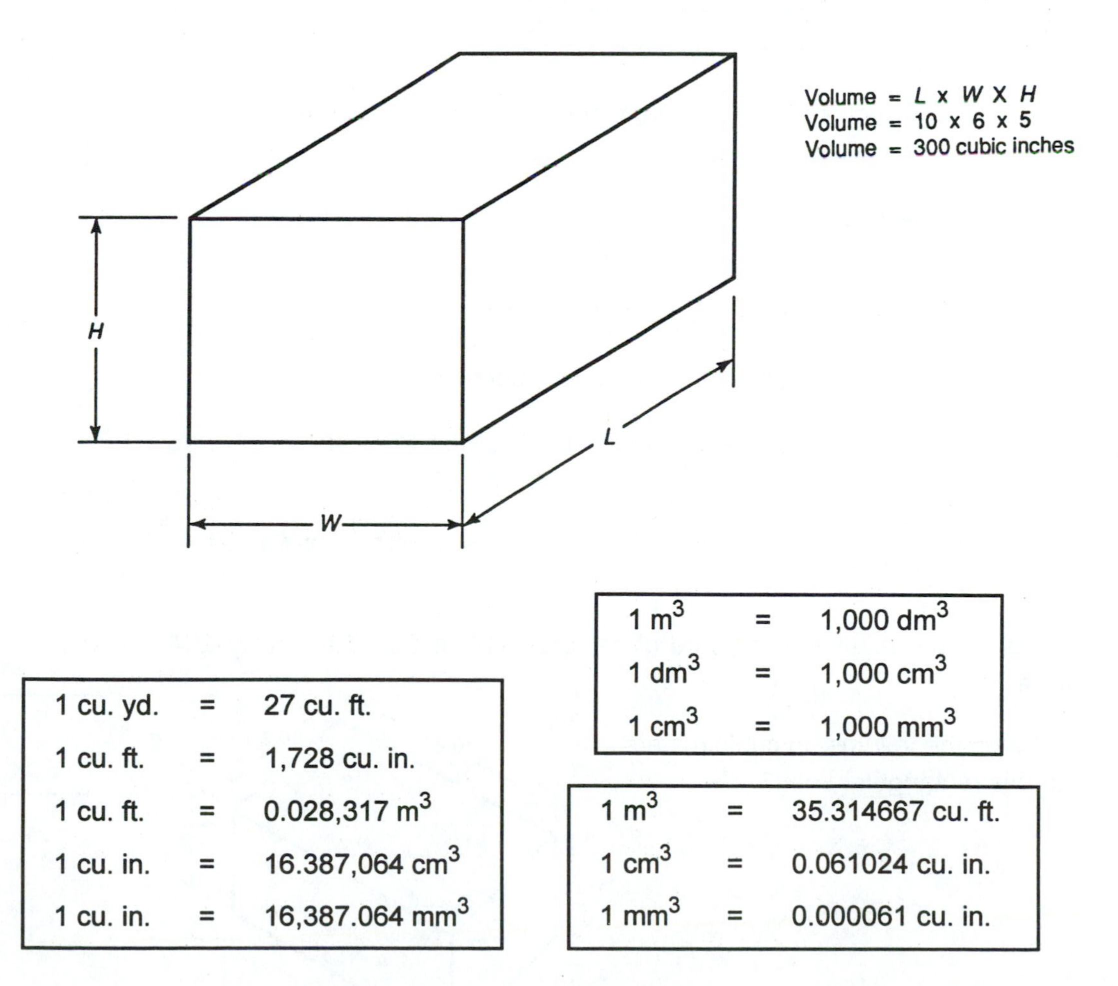

1 m^3	=	1,000 dm^3
1 dm^3	=	1,000 cm^3
1 cm^3	=	1,000 mm^3

1 cu. yd.	=	27 cu. ft.
1 cu. ft.	=	1,728 cu. in.
1 cu. ft.	=	0.028,317 m^3
1 cu. in.	=	16.387,064 cm^3
1 cu. in.	=	16,387.064 mm^3

1 m^3	=	35.314667 cu. ft.
1 cm^3	=	0.061024 cu. in.
1 mm^3	=	0.000061 cu. in.

PRACTICAL PROBLEMS

Use this diagram for problems 1–3.

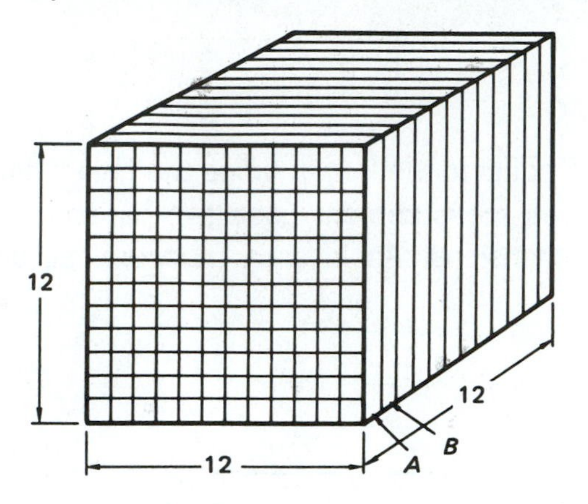

1. Slice *A*, one inch thick, is cut from the cube. How many 1-inch cube blocks are in this slice? __________

2. How many cubic inches are contained in a 3-inch-thick piece cut from this cube? __________

3. How many cubic inches are in the entire cube? __________

4. What is the volume, in cubic inches, of a 1-foot cube? __________

5. Find the volume of a piece of $\frac{3}{4}$-inch by $1\frac{1}{2}$-inch steel bar which is 6 inches long. __________

6. Find the difference, in cubic inches, between 3 cubic feet and a 3-foot cube. __________

7. What is the volume of a piece of flat stock 19 mm by 50 mm by 254 mm? __________

8. What is the volume, in cubic meters, of this rectangular solid? __________

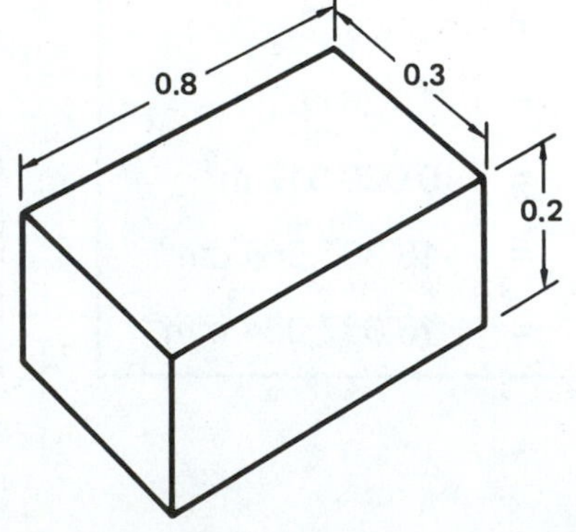

9. Find, in cubic millimeters, the volume of each rectangular solid.

	A	B	C	VOLUME
a.	30 mm	100 mm	450 mm	
b.	15 mm	0.01 m	2.3 m	
c.	25 mm	75 mm	1 m	
d.	1 cm	3 cm	1.7 dm	
e.	1 in.	3 in.	12 in.	

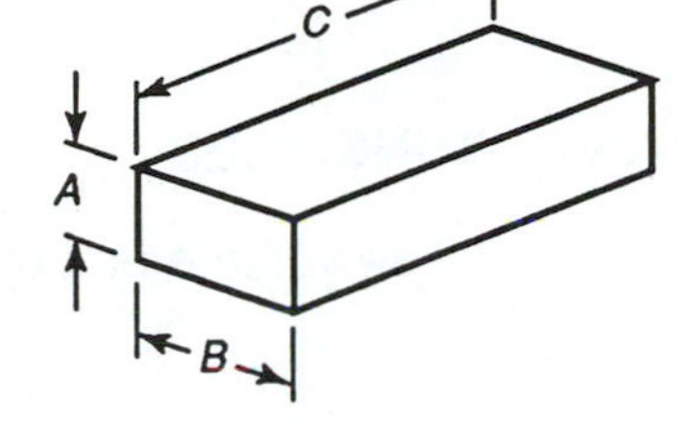

10. This cast iron sleeve has a cored hole that extends through its entire length. The hole is 25 mm wide and 100 mm high. How many cubic millimeters of iron are there in this sleeve? __________

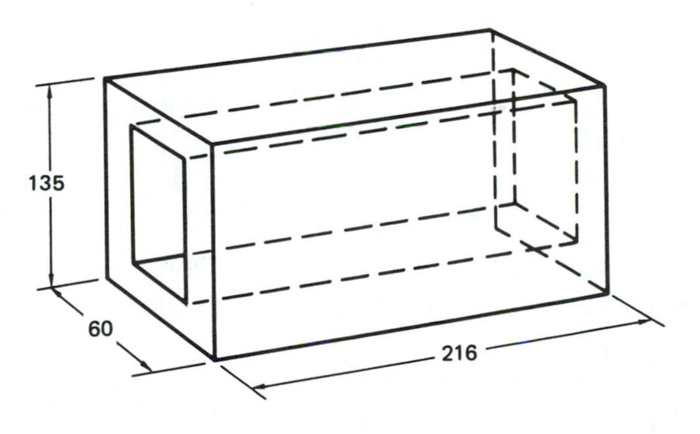

11. How many cubic inches of metal are removed by cutting a square blank in this hexagonal part? Round the answer to four decimal places. __________

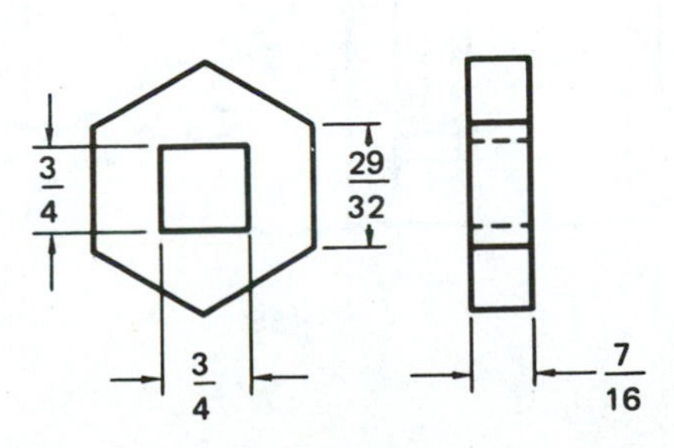

Unit 25 VOLUME OF CYLINDRICAL SOLIDS

BASIC PRINCIPLES

- The volume of a cylinder may be calculated using the following formula:

$$\text{Volume} = \pi r^2 h$$

This formula is the same as the circular area formula, except we have now added the third dimension, height. As in other volume measurements, volume is expressed in *cubic* units.

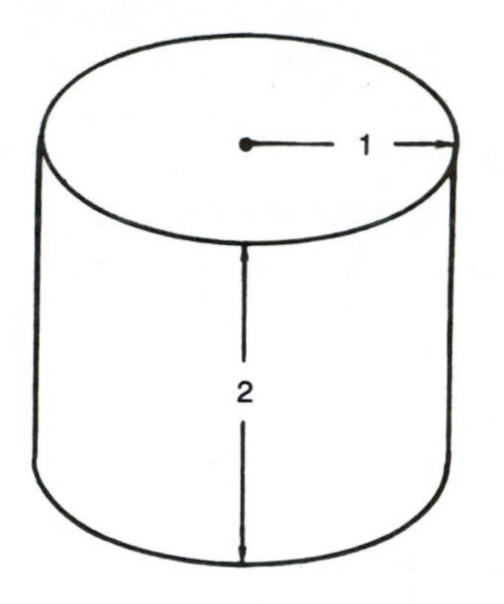

Example:

Volume = $\pi r^2 h$
Volume = $3.1416 \times 1 \times 2$
Volume = 6.2832 cubic inches

PRACTICAL PROBLEMS

1. Calculate the volume of material removed when boring a 50-mm-diameter hole through a flat piece of steel that is 35 mm thick. __________

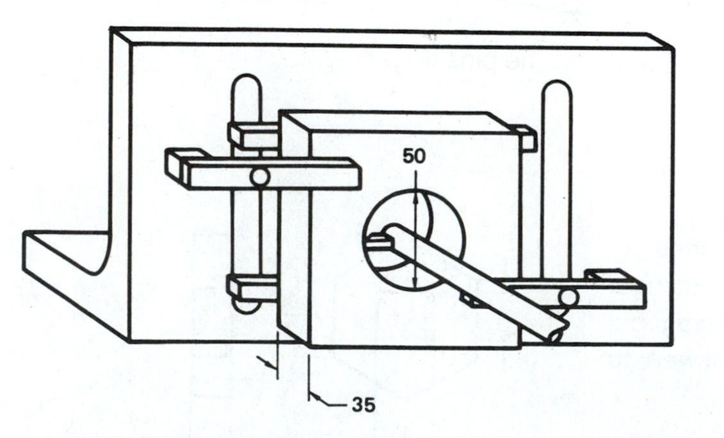

2. What is the volume of this cast iron sleeve in cubic inches? ________

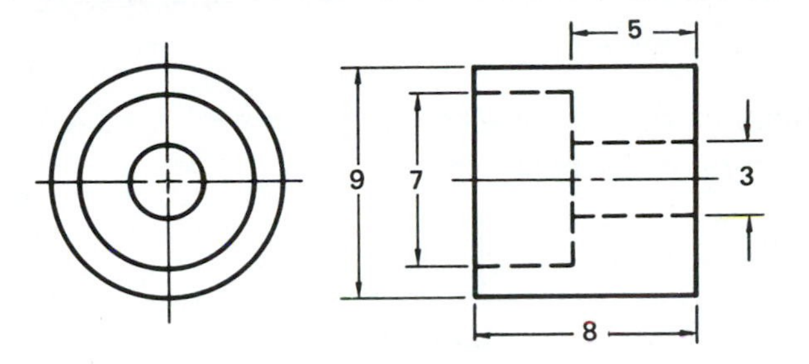

3. Find the volume of the bushing shown. Express the answer to four decimal places. ________

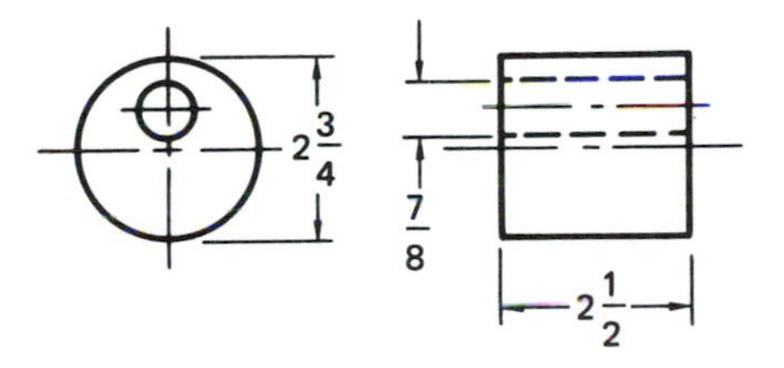

4. Find the volume of a piece of 50-mm-diameter rod that is 300 mm long. ________

5. The volume of a 3.2-cm-diameter cylinder is 98.52 cubic centimeters. Find the length to the nearer hundredth centimeter. ________

6. Find the length of a $1\frac{1}{8}$-inch-diameter cylinder with a volume of 5.467 cubic inches. Express the answer to the nearer thousandth inch. ________

7. Calculate the total volume of 408 of the pins shown. ________

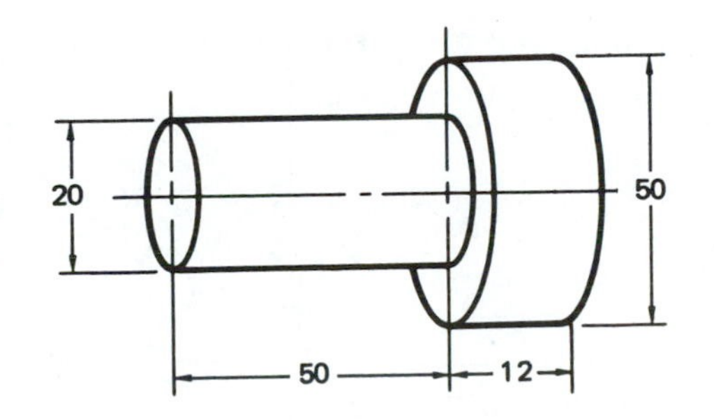

8. A copper tube has an outside diameter of 2 inches and an inside diameter of $1\frac{1}{2}$ inches. Determine the number of cubic inches of metal in one foot of length. ________

9. Find, to the nearer ten-thousandth cubic meter, the volume of a steel shaft 6 centimeters in diameter and 20 meters long. ________

Note: Use this diagram for problems 10–12.

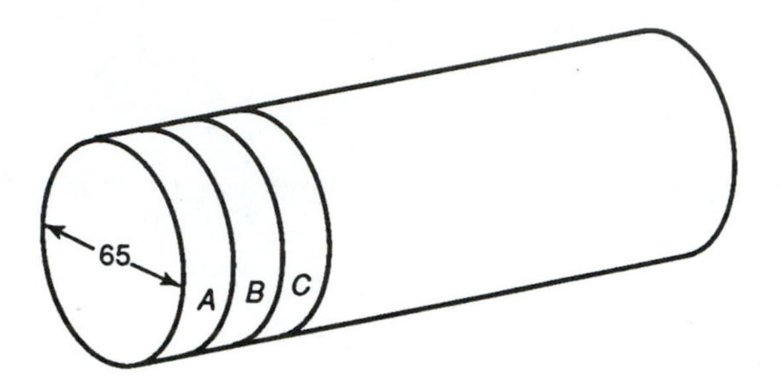

10. Slice *A*, 25 mm long, is cut from the cylinder. How many cubic millimeters are in this slice? ________

11. Slices *A*, *B*, and *C*, each 25 mm long, are cut from the cylinder. How many cubic millimeters do the three slices contain? ________

12. Find, in cubic millimeters, the volume of the piece of this rod that is 250 mm long. ________

13. A piece of stock is turned on a lathe from $2\frac{1}{4}$ inches in diameter down to $1\frac{3}{4}$ inches in diameter. If the cut is 7 inches long, how many cubic inches of metal are removed? ________

14. A $\frac{3}{4}$-inch diameter flat bottom drill is used to make a hole $2\frac{1}{4}$ inches deep. How many cubic inches of metal are removed? Round the answer to the nearer thousandth cubic inch. ________

Unit 26 MASS (WEIGHT) MEASURE

BASIC PRINCIPLES

Example:

A 1-inch-square steel bar weighs 3.396 lb./ft. How much does a 20-foot bar of this material weigh?

Weight = pounds per foot x length in feet
Weight = 3.396 x 20
Weight = 67.92 pounds

PRACTICAL PROBLEMS

1. Number *00* United States Standard Gauge steel plate weighs 14.02 pounds per square foot. Find, to the nearer thousandth, the weight of a number *00* plate that is 2 feet 7 inches by 3 feet 11 inches. __________

2. The dimensions of four triangular steel plates are given in this table. Find each weight (mass) to the nearer thousandth.

	BASE	ALTITUDE	WEIGHT (MASS) PER UNIT	TOTAL WEIGHT (MASS)
a.	14.27 in.	8.6 in.	0.175 lb. per sq. in.	
b.	0.282 m	0.484 m	0.095 kg/m^2	
c.	210.57 mm	105.4 mm	0.217 5 kg/mm^2	
d.	125 in.	$47 \frac{5}{16}$ in.	0.123 lb. per sq. in.	

3. A 4.75-mm-thick plate weight 0.024 kg/mm^2. What is the mass of a triangular plate with a base of 141.3 mm and an altitude of 117.5 mm? __________

4. A round steel bar will support a load of 40,000 pounds per square inch of its cross-sectional area. If the bar is ⅞ inch in diameter, how many pounds will it support? __________

5. Find the mass of a piece of steel 15 mm x 60 mm x 1.5 m. The steel weighs 0.14 kg/cm^3. __________

6. Using 0.283 pound per cubic inch, find the weight of a ¾-inch round bar that is 20 feet long. Round the answer to four decimal places. __________

Unit 27 VOLUME OF FLUIDS

BASIC PRINCIPLES

Example:

How many gallons of water will a tank that measures 6 feet in diameter by 10 feet long hold?

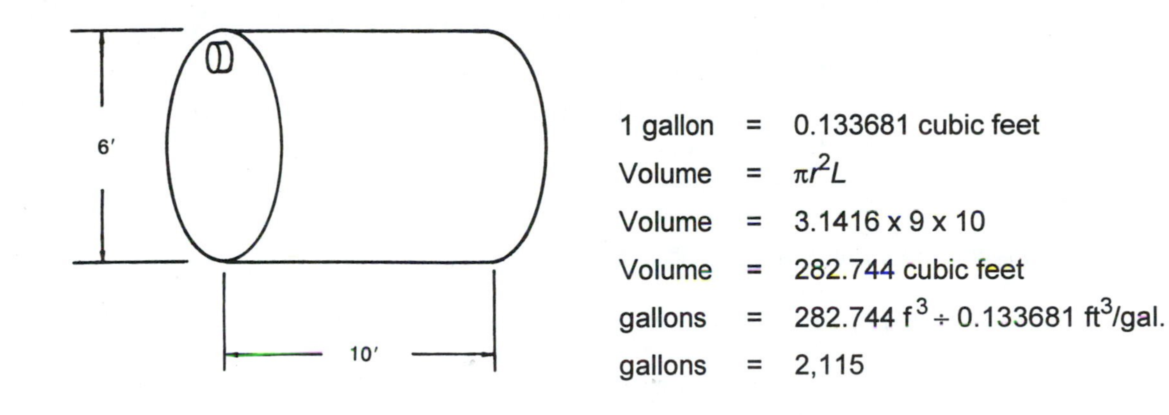

1 gallon = 0.133681 cubic feet

Volume = $\pi r^2 L$

Volume = 3.1416 x 9 x 10

Volume = 282.744 cubic feet

gallons = 282.744 $f^3 \div$ 0.133681 ft^3/gal.

gallons = 2,115

1 gallon (gal.)	=	4 quarts (qt.)
1 quart (qt.)	=	2 pints (pt.)
1 gallon	=	0.133681 cubic feet = 231 cubic inches
1 gallon	=	3.785,411 liters

1 liter (L)	=	1,000 milliliters (mL)
1 liter (L)	=	1 cubic deciliter
1 liter (L)	=	0.264172 gallon
1 liter (L)	=	1.056688 quarts

PRACTICAL PROBLEMS

1. What decimal part of a gallon is 3 liters? __________

2. A tank holding 350 gallons has two pipes connected to it. One pipe empties out 6 gallons per minute and the other empties out 17 gallons per minute. How many minutes and seconds will it take to empty the tank with both pipes flowing? Round the answer to the nearer whole second. __________

3. How many gallons of coolant will a 7.2-cubic-foot tank hold? Round the answer to the nearer hundredth gallon. ________

4. A machinist must know how many liters of oil to order to fill this tank. The inside dimensions of this tank are in decimeters.

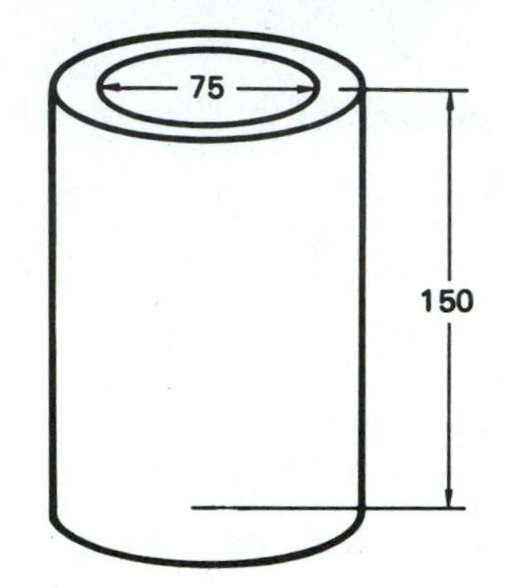

   a. What is the measure of the tank in cubic decimeters? a. ________

   b. If the tank is empty, how many liters must be ordered to fill the tank? b. ________

   c. If the oil measures to the 10-dm mark on the side of the tank, how many liters must be ordered? c. ________

5. The inside dimensions of this tank are in inches. Find, to the nearer thousandth gallon, the amount the tank will hold. ________

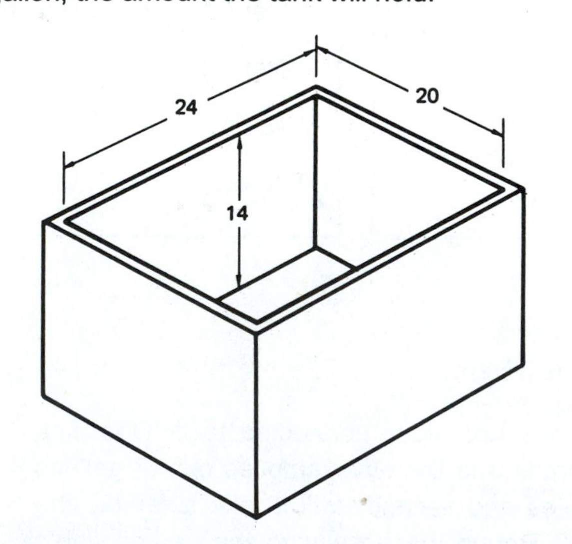

6. The inside dimensions of this tank are in meters. How many liters will this tank hold? __________

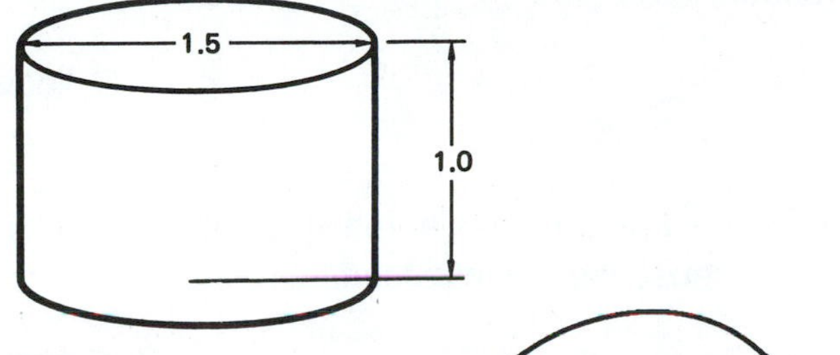

CRITICAL THINKING PROBLEMS

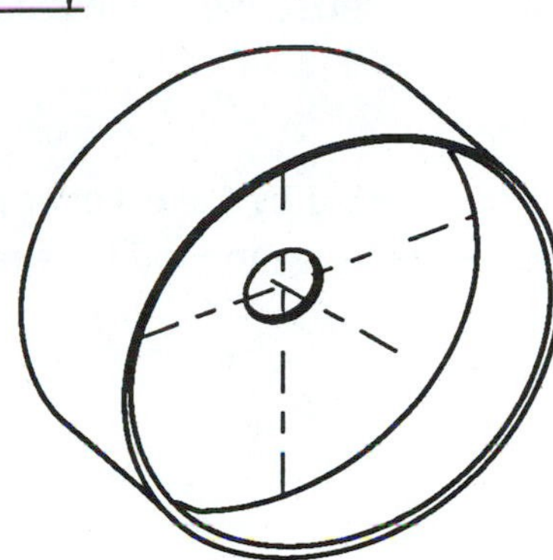

1. Determine how much paint will be required to paint the pulley shown. The width of the face is 12″ and the outside diameter is 36″. The center and the outer loop are both ½″ thick. The center hole is 6″ in diameter.

   Hint: Most paint cans will give coverage information in the instructions. __________

2. Determine the area of the piston in the hydraulic cylinder shown.

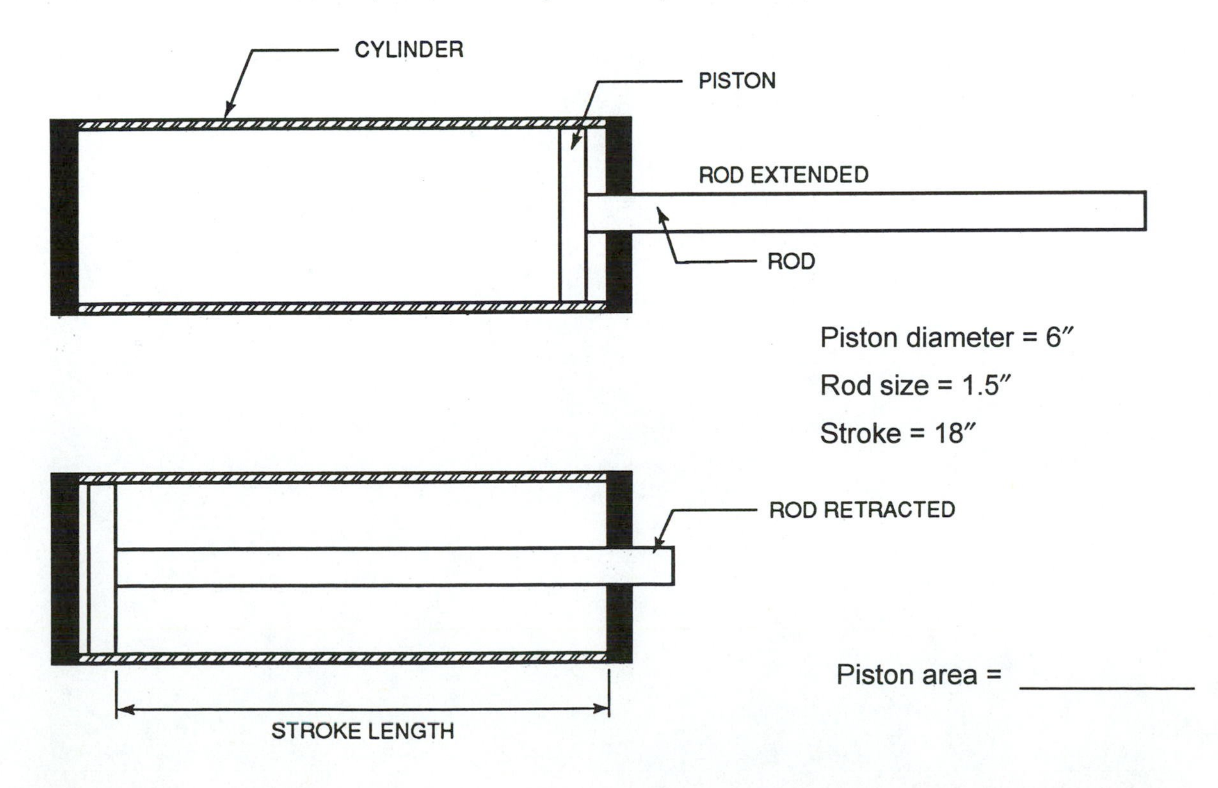

Piston diameter = 6″

Rod size = 1.5″

Stroke = 18″

Piston area = __________

3. Determine the volume of the cylinder shown in question 2. when the piston is fully extended.

Cylinder volume = __________

4. What force will be exerted by the cylinder rod when 600 PSI of hydraulic pressure is applied to cause the rod to extend.

Extending force = __________

5. What force will be exerted by the cylinder rod when 600 PSI of hydraulic pressure is applied to cause the rod to retract?

Retracting force = __________

Percent and Graphs

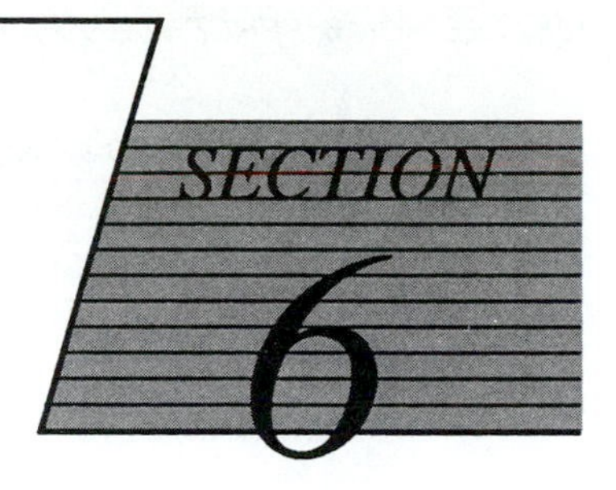

Unit 28 PERCENT

BASIC PRINCIPLES

- Percents are used to indicate some portion of a whole. For example, if there are one hundred apples in a basket, then 100% of the applies are in the basket. If one half of the apples are removed, then 50% are taken out and 50% remain in the basket.

  Example: There were 2,500 drill bits in stock before 100 were sold. What percent were sold, and what percent of the original total remain?

  Step 1: Divide the number sold by the original number on the shelf.

  $$100 \div 2{,}500 = 0.04$$

  Step 2: Multiply the quotient by 100.

  $$0.04 \times 100 = 4\% \text{ bits sold}$$

  Step 3: Subtract the product from 100%.

  $$100\% - 4\% = 96\% \text{ bits remaining}$$

  It can be seen that there are 2,400 bits left in stock after the sale, This could have been determined by the following method:

  Step 1: Divide 96% by 100 to get the percent back to decimal form.

  $$96\% \div 100 = 0.96$$

  Step 2: Multiply 2,500 by 0.96.

  $$2500 \times 0.96 = 2{,}400 \text{ bits remaining}$$

Remember to convert a percent to a decimal when using the percent in a calculation.

- To convert a fraction to a percent convert the fraction to a decimal fraction and multiply by 100.
- **Example:** Convert $^3/_4$ to a percent.

$$\frac{3}{4} = 0.75$$
$$0.75 \times 100 = 75\%$$

- To convert a percent to a fraction, divide the percent by 100 and reduce it to lowest terms.

Example: Change 25% to a fraction.

$$25 \div 100 = \frac{25}{100} = \frac{1}{4}$$

PRACTICAL PROBLEMS

1. Express each decimal fraction as a percent.

   a. 0.85 — a. ____________

   b. 0.03 — b. ____________

   c. 0.15 — c. ____________

2. Express each percent as a decimal fraction.

   a. 20 percent — a. ____________

   b. 98.9 percent — b. ____________

   c. 115 percent — c. ____________

3. Express each percent as a common fraction.

   a. 40 percent — a. ____________

   b. $66^2/_3$ percent — b. ____________

   c. 75 percent — c. ____________

4. Express each common fraction as a percent.

   a. $\frac{3}{10}$ a. ___________

   b. $\frac{5}{8}$ b. ___________

   c. $\frac{1}{10}$ c. ___________

5. Out of a lot of 60 castings, 5 percent are rejected because of defects. How many are rejected? ___________

6. If $3\frac{1}{3}$ percent of a lot of 60 castings are defective, how many are rejected? ___________

7. Of a lot of 72 castings, $12\frac{1}{2}$ percent are scrapped. How many rejects does this represent? ___________

8. An inspection of 96 castings reveals $16\frac{2}{3}$ percent to be defective. How many failed to pass inspection? ___________

9. A common fluid used to etch metal is made from 1 part nitric acid and 4 parts water.

   a. What percent is nitric acid? a. ___________

   b. What percent is water? b. ___________

10. A shop supervisor has to determine the percent of work spoiled per job. The percent scrapped is $\frac{\textit{spoiled pieces}}{\textit{number of workpieces}} \times 100$. Using the listed information, calculate the percent of scrapped work to the nearer hundredth percent.

	NUMBER OF WORKPIECES	NUMBER OF SPOILED PIECES	PERCENT SCRAPPED
a.	28	6	
b.	23	11	
c.	50	4	

11. Find the percent of metal removed from a machined casting that weighed 12 pounds before machining and weighed 9.25 pounds after machining. Express the answer to the nearest hundredth. ________

12. Determine the percent of metal removed in a machining operation if the weight of the item was 14.75 pounds before machining and weighs 12.37 pounds after machining. Round the answer to the nearest hundredth. ________

Note: Use this information for problems 13–16.

$$\text{percent efficiency} = \frac{\text{output}}{\text{input}} \times 100$$

or

$$\text{percent efficiency} = \frac{\text{power delivered}}{\text{power supplied}} \times 100$$

$$\text{output} = \text{input} \times \text{percent efficiency}$$

or

$$\text{power delivered} = \text{power supplied} \times \text{percent efficiency}$$

$$\text{input} = \frac{\text{output}}{\text{percent efficiency}}$$

or

$$\text{power supplied} = \frac{\text{power delivered}}{\text{percent efficiency}}$$

$$\text{percent of loss} = \frac{\text{loss}}{\text{input}} \times 100$$

$$\text{input} = \frac{\text{loss}}{\text{percent of loss}}$$

$$\text{percent of loss} + \text{percent efficiency} = 100\%$$

13. A 5-horsepower motor running at rated capacity is attached to a machine that delivers 4.2 horsepower. What is the machine efficiency? ________

14. In a gear train, 4 percent of the power supplied is lost in friction. The power loss is found to be $2\frac{1}{2}$ horsepower.

    a. What is the power supplied? a. ___________

    b. What is the power delivered? b. ___________

15. The belt drive on a grinder has a loss of $\frac{2}{3}$ horsepower with 97 percent efficiency. Round each answer to the nearer hundredth.

    a. What is the power supplied? a. ___________

    b. What is the power delivered? b. ___________

16. A drive through belts and countershafts is found to have 76 percent efficiency. The loss is 2.314 horsepower. Find, to the nearer thousandth, the power delivered to the machine. ___________

17. Calculate the total weight of each casting lot. Round to the nearer hundredth, if necessary.

	Percent Rejected	Weight Rejected	Total Weight
a.	$6\frac{1}{2}$ percent	19 pounds	
b.	$17\frac{3}{4}$ percent	37 pounds	
c.	3 percent	30 pounds	

Unit 29 INTERPRETING GRAPHS

BASIC PRINCIPLES

- *Graphs* are used as a visual aid to understanding and comparing quantities. Graphical format makes large groups of numbers easier to understand.

PRACTICAL PROBLEMS

Note: Use this graph and information for problems 1–5.

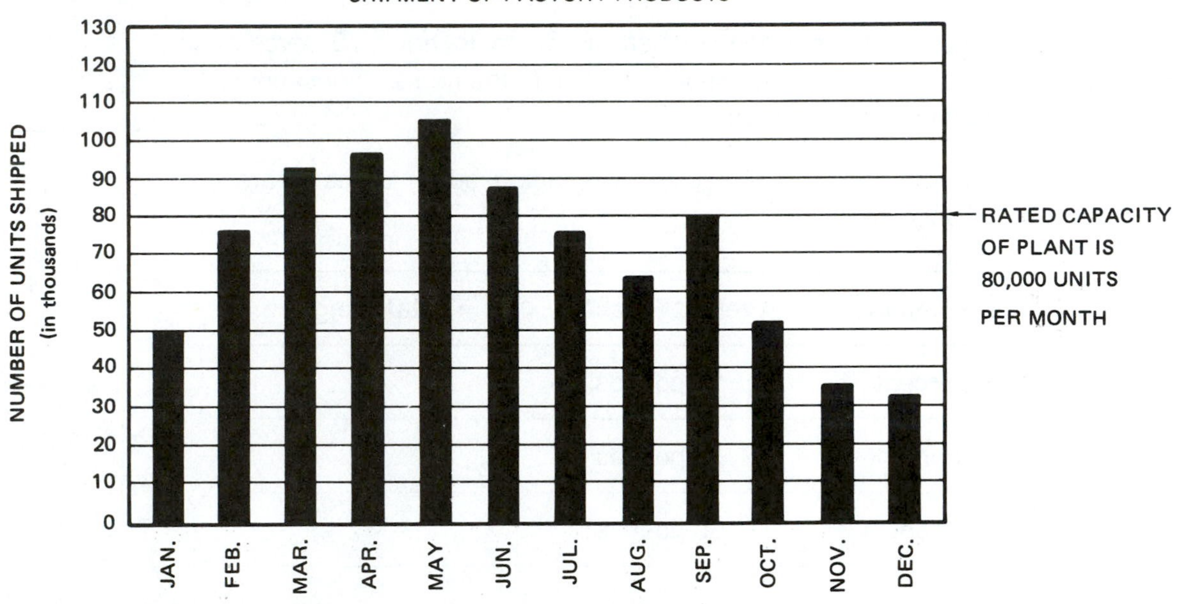

Using this graph determine whether each statement is true or false. Use ***T*** for true and ***F*** for false.

1. More units are shipped in May than any other month. __________
2. The plant was working above rated capacity for one-half of the year. __________
3. The average for the four high months is above 90,000 units per month. __________
4. The average throughout the year is above the rated capacity of the plant. __________
5. The peak load is in the month of May. __________

Note: Use this graph and information for problems 6–8.

A line graph can be used to estimate unknown values. This line graph shows temperature readings from midnight to noon in degrees Fahrenheit (°*F*) and degrees Celsius (°*C*).

The equivalences between these readings are:

degrees Celsius (°*C*) = (°*F* - 32°) X 0.5556

degrees Fahrenheit (°*F*) = (°*C* X 1.8) + 32°

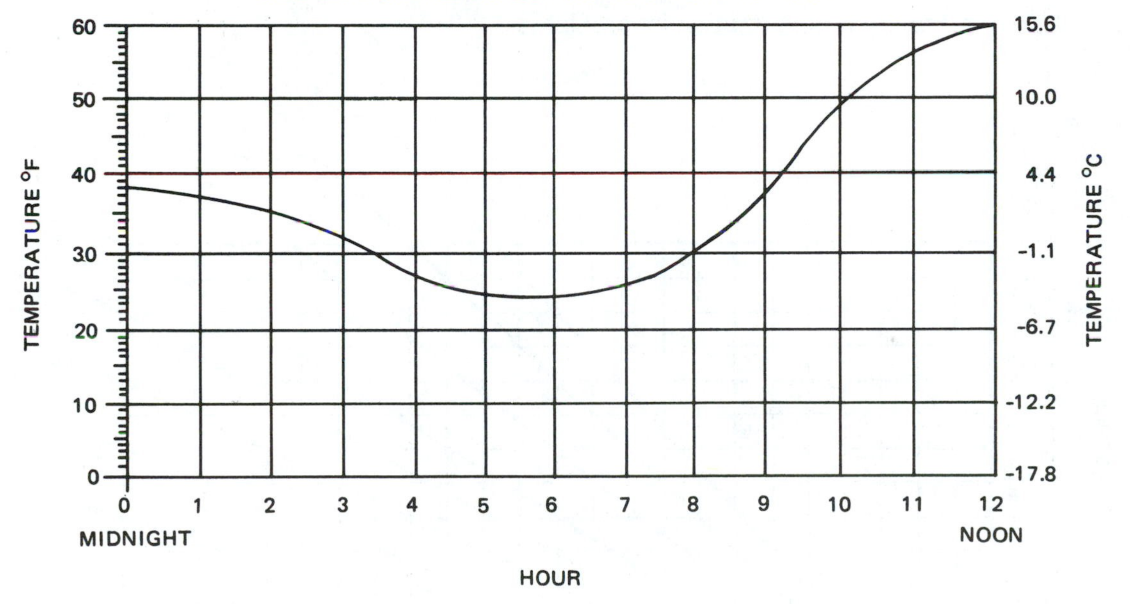

To estimate the temperature at 8:30, draw a vertical line midway between 8 :00 and 9:00. Draw a horizontal line from the point where the curve intersects the line for 8:30 to the temperature scale. This gives a temperature reading of 34 °*F* or 1.1 °*C*.

Using this chart find the temperature readings (°*F* and °*C*) for each time. Express Celsius-degree temperatures to the nearer tenth.

6. 1:00 A.M.

   a. degrees Fahrenheit a. ____________

   b. degrees Celsius b. ____________

7. 2:30 A.M.

   a. degrees Fahrenheit a. ____________

   b. degrees Celsius b. ____________

8. 9:30 A.M.

   a. degrees Fahrenheit a. ____________

   b. degrees Celsius b. ____________

Note: Use this graph and information for problems 9–13.

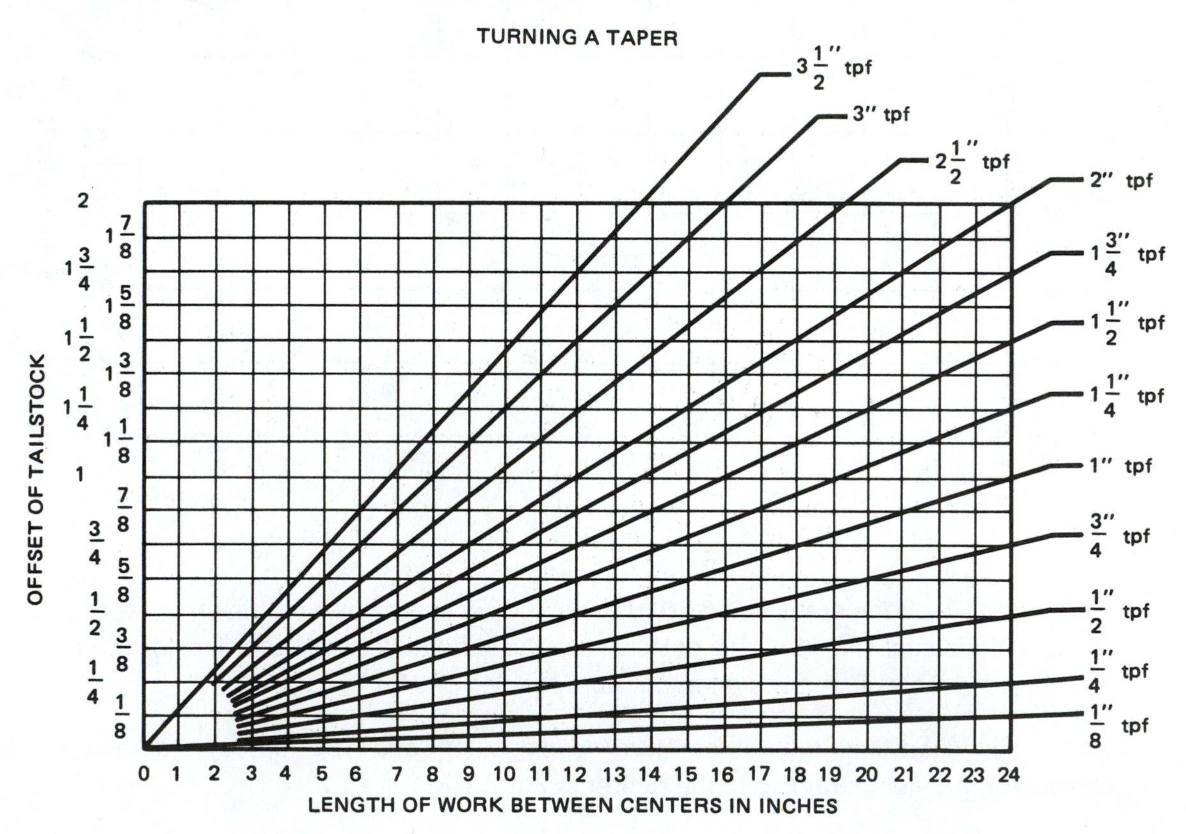

This graph may be used for determining the amount to set over the tailstock of a lathe for turning a taper. The length of the work and the taper per inch are known. The horizontal lines represent the offset. The amount of offset is shown in the column of dimensions at the left. The oblique lines represent taper per foot. This amount is marked on the line at its extreme right.

Using this graph, determine each missing factor for turning each of these tapers.

	Length of Work	Taper per Foot	Offset
9.	10 inches	$5/8$ inch	________
10.	8 inches	$3/4$ inch	________
11.	14 inches	________	$5/8$ inch
12.	4 inches	________	$5/8$ inch
13.	________	$3/8$ inch	$1/4$ inch

CRITICAL THINKING PROBLEMS

1. A certain paint has the following percentages by weight of solvents.

   Toluene 23%

   Xylene 11%

   Ethyl benzene 1.9%

   Determine the pounds of each solvent in a gallon of the paint. The paint weighs 10.3 pounds per gallon. ________

2. If 4,623 gallons of the paint in question #1 were used during the year, how many tons of each solvent were used? ________

3. During the next year the company using the paint above switched to a high-solids paint. The high-solids paint has no toluene, 35% less xylene, and 5% less ethyl benzene than the previous paint. Determine the tons of each solvent used if 6,989 gallons of the high-solids paint were used. ________

Ratio and Proportion

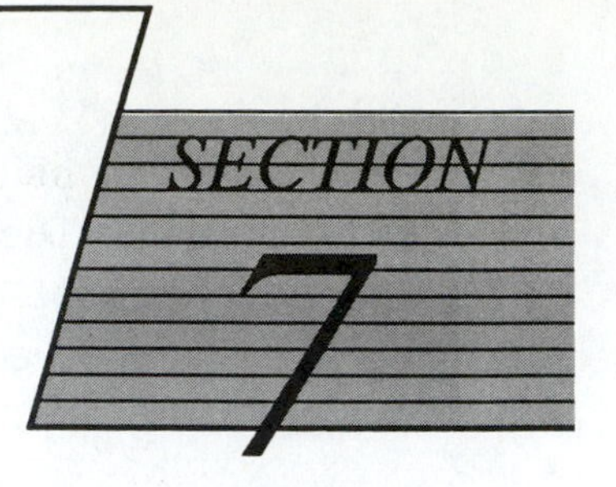

Unit 30 RATIO

BASIC PRINCIPLES

- A *ratio* is a comparison of one quantity to another quantity. Ratios are used quite often when referring to sets of gears. A *gear ratio* is a comparison of the numbers of teeth on two gears, reduced to lowest terms. Usually a gear ratio expresses the driven gear first and the driver next.

Example:

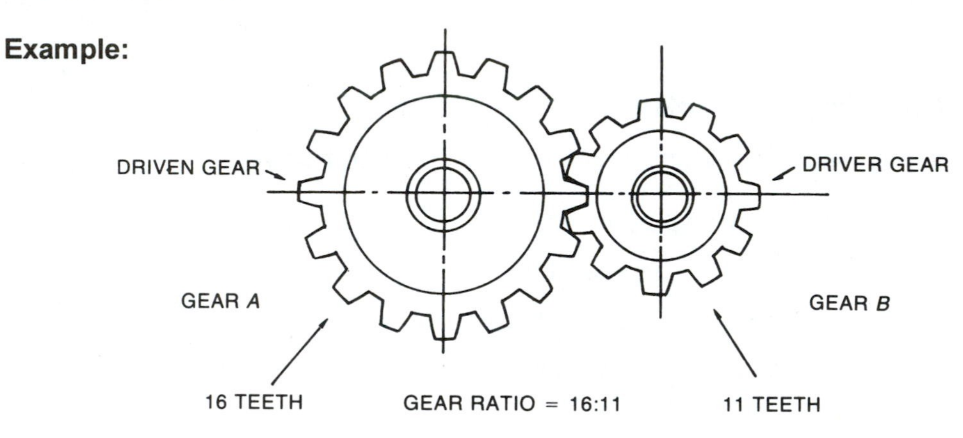

This ratio would normally be reduced to 1.45:1.

PRACTICAL PROBLEMS

1. Gear *A* has 80 teeth and gear *B* has 40 teeth.

   a. What is the ratio of *A* to *B*?

   b. What is the ratio of *B* to *A*?

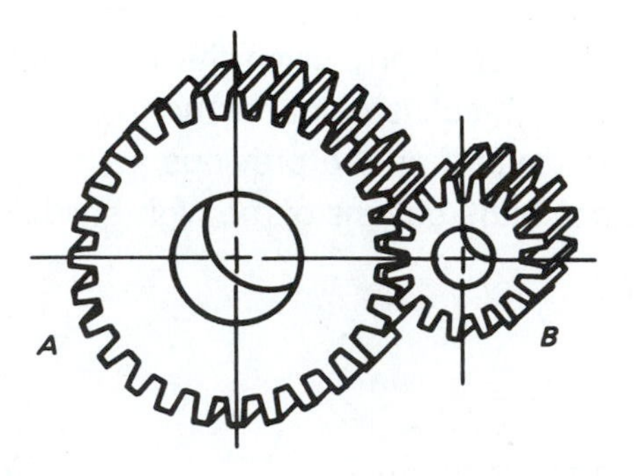

a. ____________

b. ____________

2. Two gears, *A* and *B* are working together. *A* has 60 teeth and *B* has 20.

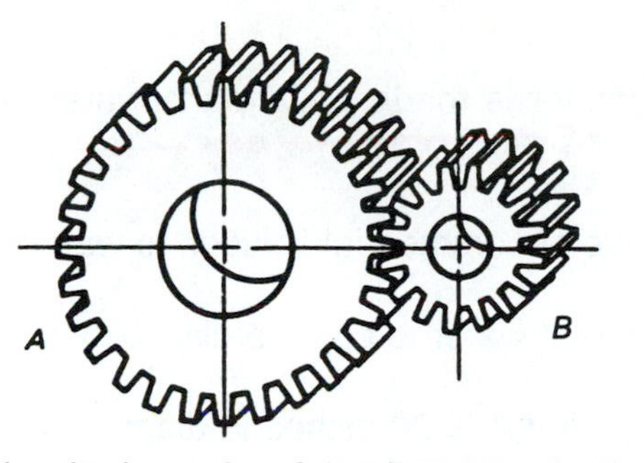

   a. What is the ratio of *A* to *B*? a. ____________

   b. What is the ratio of *B* to *A*? b. ____________

3. If two gears each have 48 teeth, what is the ratio of the first gear to the second? ____________

4. A certain alloy is made with 15 kg of metal *A* and 45 kg of metal *B*.

   a. What is the ratio of *A* to *B*? a. ____________

   b. What is the ratio of *B* to *A*? b. ____________

Note: Use this chart for problems 5–10.

MATERIAL	APPROXIMATE MASS (in kg/m^3)
Steel, Carbon	7,800
Aluminum	2,600
Bismuth	9,600
Tin	7,200
Brass	8,000

5. What is the ratio of steel to tin? ____________

6. What is the ratio of bismuth to brass? ____________

7. What is the ratio of steel to aluminum? ____________

8. What is the ratio of brass to aluminum? ____________

9. What is the ratio of bismuth to steel? ____________

10. What is the ratio of aluminum to bismuth? ____________

11. A coolant for a grinder is made of 25 parts water to 1 part commercial solution.

    a. What is the ratio of commercial solution to water? a. ___________

    b. What is the ratio of water to commercial solution? b. ___________

12. In this belt drive, pulley *A* is 20 inches in diameter; pulley *B* is 10 inches in diameter. What is the ratio of the diameter of *A* to the diameter of *B*? ___________

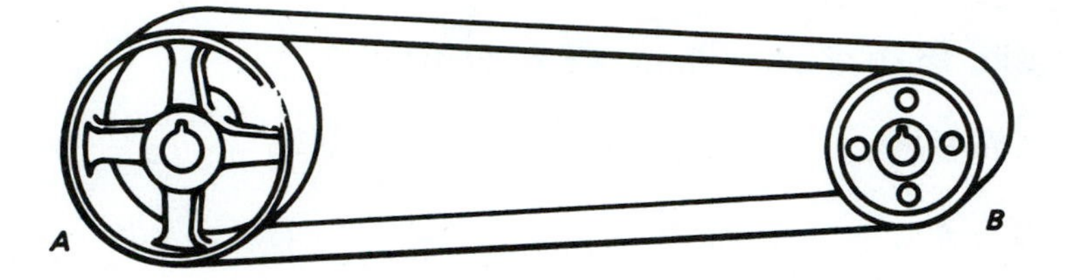

Unit 31 DIRECT PROPORTION

BASIC PRINCIPLES

- A *proportion* is two ratios that are equal.

Example:

$$3:1 = 24:8$$

This can also be written as:

$$\frac{3}{1} = \frac{24}{8}$$

Direct proportion can be used to solve problems concerning ratios.

Example:

If a worker makes $16.00 in 3 hours, how much will he make in 33 hours?

Step 1: Set up the proportion

$$\frac{3\text{hr}}{33\text{hr}} = \frac{\$16.00}{x}$$

Step 2: Cross multiply.

$$\frac{3\text{hr}}{33\text{hr}} = \frac{\$16.00}{x}$$

$$3x = 528$$

$$x = \$176$$

PRACTICAL PROBLEMS

1. A shop employing 15 people is able to build 12 machines in one month. If the production schedule calls for 20 machines in one month, how many people are required? ____________

2. The sides of similar triangles form these proportions.

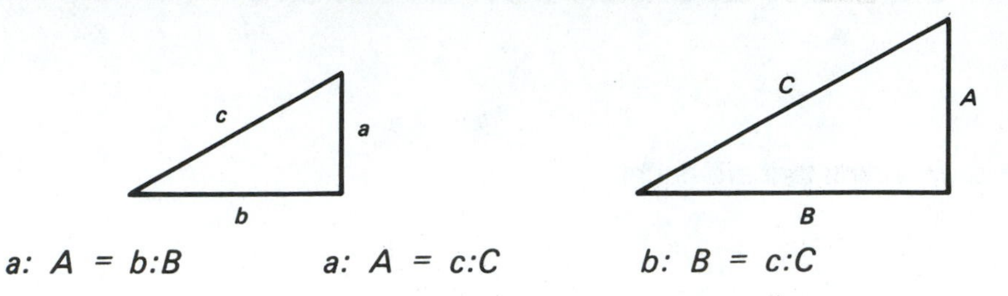

a: A = b:B   a: A = c:C   b: B = c:C

Using these proportions, find each value to the nearer thousandth.

	a	A	b	B	c	C
a.	6	20	8		10	
b.	7	11	3		9	
c.		19	5	12		27

3. A piece tapers 0.796 inch in $2\frac{1}{2}$ inches of length. What length must the piece be to taper 1 inch? Express the answer to the nearer thousandth inch. __________

4. The taper of a piece is $\frac{1}{2}$ inch per foot. What is the length of the piece when the taper is 0.0359 inch? Express the answer to four decimal places. __________

5. A square bar of steel 3.47 meters long weighs 21.072 kilograms. What is the mass of a piece of the same type of bar 2.927 meters long? Round the answer to the nearer thousandth. __________

6. A company must now produce a part that is 57.2 mm long and is 1.25 times larger than the original part. How large is the original part? __________

7. To express degrees as minutes, this proportion may be used.

$$\frac{1 \text{ degree}}{60 \text{ minutes}} = \frac{? \text{ degrees}}{x \text{ minutes}}$$

Use this proportion to find each value to the nearer minute.

	DEGREES	MINUTES
a.	0.94°	
b.	0.42°	
c.	0.50°	
d.	0.30°	
e.	0.19°	

Note: Use this information for problems 8 and 9.

To find the number of turns to index divisions, this proportion is used.

$$\frac{40}{\text{number of required turns}} = \frac{\text{number of divisions}}{1}$$

The number 40 is the total number of turns for the spindle to revolve once and 1 represents the whole circle.

For example, to make 5 divisions, the proportion is:

$$\frac{40}{x} = \frac{5}{1} \quad \textit{or} \quad x = \text{8 turns of the hand crank}$$

An indexing plate is used to accurately divide fractional portions of a turn. The plates have a series of holes and are available in many sizes.

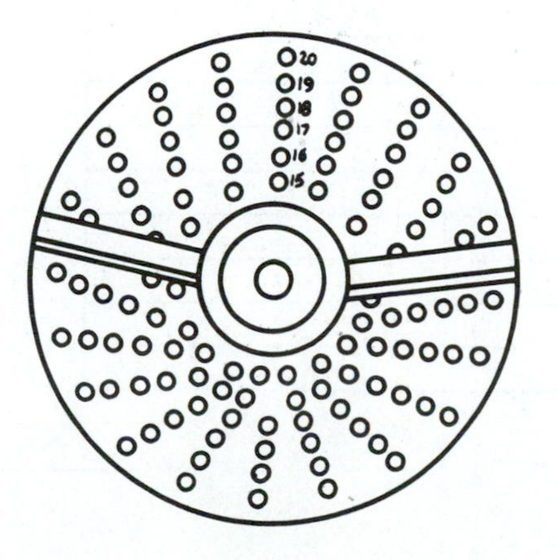

This chart shows the number of holes for three plates.

PLATE # 1	PLATE # 2	PLATE # 3
15 holes	21 holes	37 holes
16 holes	23 holes	39 holes
17 holes	27 holes	41 holes
18 holes	29 holes	43 holes
19 holes	31 holes	47 holes
20 holes	33 holes	49 holes

When the number of turns is a whole number, any plate and any number of holes may be used. When the number of turns contains a fraction, the plate and the number of holes to be used depends on the fractional portion of the turn. For example, to index 24 divisions, $1\frac{2}{3}$ turns of the hand crank are required. To turn $\frac{2}{3}$ turn, select a plate in which the number of holes is a multiple of the denominator 3. Plates with 15, 18, 21, 27, 33, and 39 holes could be used. The plate used should be the one with the least number of holes.

The 15-hole plate is selected. Once the place is selected, this direct proportion is used.

$$\frac{2}{3} = \frac{x}{15} \text{ or } x = 10$$

In this case, 10 holes in a 15-hole indexing plate will accurately divide the required $\frac{2}{3}$ turn.

8. Complete this chart of the required number of turns to index these divisions. Express the fractions in lowest terms.

	NUMBER OF DIVISIONS	NUMBER OF TURNS OF THE HAND CRANK
a.	9	
b.	11	
c.	4	
d.	6	
e.	18	
f.	12	
g.	3	
h.	22	
i.	2	
j.	8	

9. Complete this chart. Use the least number of holes in the indexing plate.

	DIVISIONS REQUIRED	NUMBER OF TURNS OF CRANK	NUMBER OF HOLES IN INDEXING PLATE	NUMBER OF HOLES USED
a.	3			
b.	5			
c.	6			
d.	7			
e.	11			
f.	16			
g.	19			
h.	20			
i.	21			
j.	36			
k.	45			
l.	49			
m.	54			
n.	220			
o.	376			

10. Two gears have a ratio of 3 to 1. Gear *A*, the larger of the two gears, has 33 teeth. How many teeth are there in gear *B*? ________

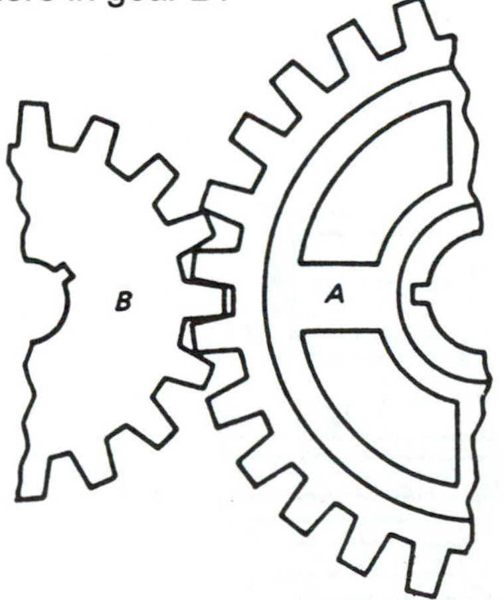

11. A shop drawing has a scale of ¼:1. The ratio ¼:1 is the ratio of the drawing to the part. How large would a line 2 inches long on the drawing actually be on the part? ________

12. Two gears in mesh have a ratio of 5 to 4. If the larger of the two gears has 60 teeth, how many teeth are there in the other gear? ________

Unit 32 INDIRECT PROPORTION

BASIC PRINCIPLES

- *Indirect proportion* is used in situations where the ratios vary inversely. An example of this would be a gear ratio where the larger gear turns slower than the smaller gear. The proportion is set up in the same manner as a direct proportion, except one side of the equation must be inverted. Once the inversion is accomplished, cross multiply as in direct proportion.

Example:

If a 100-tooth gear is turning at 100 rpm, how fast is the 35-tooth gear that is driving it turning?

Step 1: Set up proportion.

$$\frac{100 \text{ teeth}}{35 \text{ teeth}} = \frac{x}{100\ rpm}$$

Step 2: Cross multiply.

$$100\frac{\text{teeth}}{35}\text{teeth} = \frac{x}{100}\text{rpm}$$

$$35x = 10{,}000$$

$$x = 286 \text{ rpm}$$

PRACTICAL PROBLEMS

1. A step pulley drive is shown.

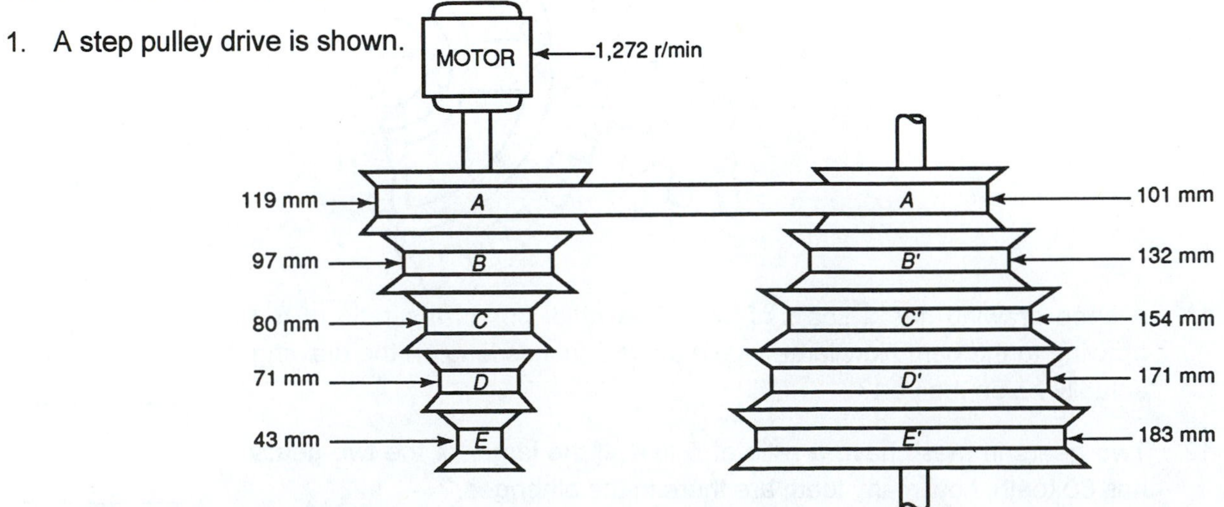

Complete this chart of diameters and speeds for each set of pulleys. Round to the nearer whole number, when necessary.

	DRIVER PULLEYS			DRIVEN PULLEYS		
	pulley	diameter	r/min	pulley	diameter	r/min
a.	*A*		1,272	*A′*		
b.	*B*		1,272	*B′*		
c.	*C*		1,272	*C′*		
d.	*D*		1,272	*D′*		
e.	*E*		1,272	*E′*		

Note: Use this diagram for problems 2–5.

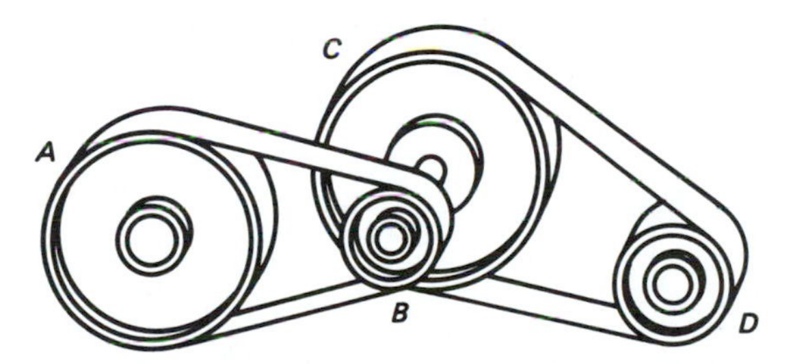

2. Pulley *A* is 406.4 mm in diameter and turns 220 revolutions per minute. Pulley *B* is 152.4 mm in diameter and pulley *C* is 609.6 mm in diameter. What diameter pulley must be placed at *D* to obtain 1,760 revolutions per minute? Express the answer to the nearer tenth. ____________

3. Pulley *B* is 762 mm in diameter, pulley *C* is 1,143 mm in diameter, and pulley *D* is 457 mm in diameter. The speed of pulley *D* is 380 revolutions per minute and the speed of pulley *A* is 420 revolutions per minute. Find the diameter of pulley *A* to the nearer tenth. ____________

4. Pulley *A* is 279 mm in diameter, pulley *B* is 914 mm in diameter, and pulley *D* is 508 mm in diameter. The speed of *A* is 840 revolutions per minute and the speed of *D* is 700 revolutions per minute. What is the diameter of pulley *C* to the nearer tenth? ____________

5. Pulley *A* is 22 inches in diameter, pulley *C* is 35 inches in diameter, and pulley *D* is 24 inches in diameter. The speed of pulleys *A* and *D* is 200 revolutions per minute. Find the diameter of pulley *B* to the nearer tenth. __________

6. A driven pulley has a 10-inch and a 12-inch step. The driving cone, running at 200 revolutions per minute, has a 6 $\frac{1}{8}$-inch step.

   a. What is the driven-cone pulley speed for the 10-inch step? a. __________

   b. What is the driven-cone pulley speed for the 12-inch step to the nearer tenth? b. __________

7. Determine the rpm of an 18-inch pulley if it is driven by a 24-inch pulley turning at 245 rpm. Express the answer to the nearer tenth. __________

8. Determine the diameter of the driving pulley on the drill press shown. Express the answer to the nearer tenth. __________

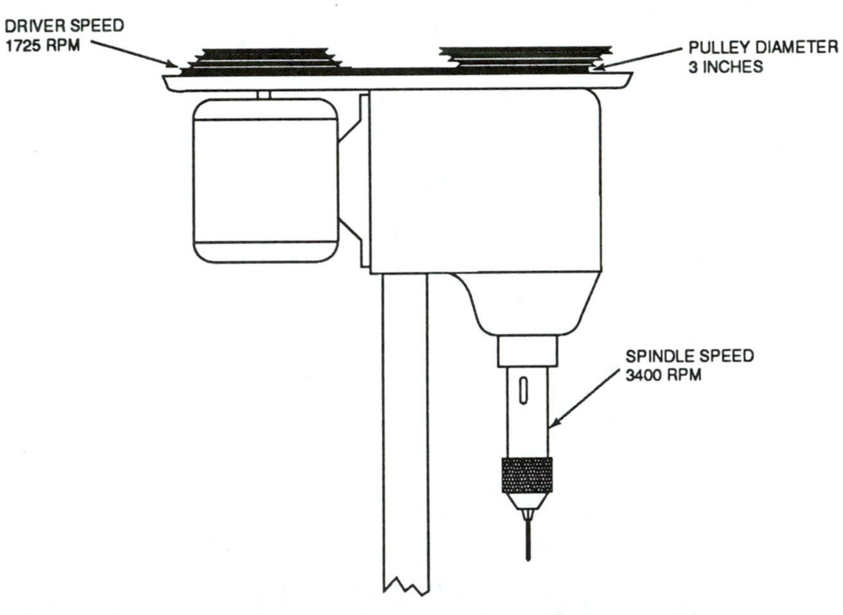

9. A 96-tooth gear is meshed with a gear having 40 teeth. If the large gear revolves 65 turns per minute, how many turns does the small gear make? __________

10. Two gears have a speed ratio of 4.6 to 1. If the smaller gear has 15 teeth, what is the number of teeth on the larger gear? __________

11. A pinion gear has 20 teeth and turns 100 revolutions per minute. To achieve 250 revolutions per minute on the driven gear, how many teeth are needed? __________

12. These four gears are in a train. A driving shaft turns at 400 revolutions per minute. Find, to the nearer whole number, the speed of the driven shaft. __________

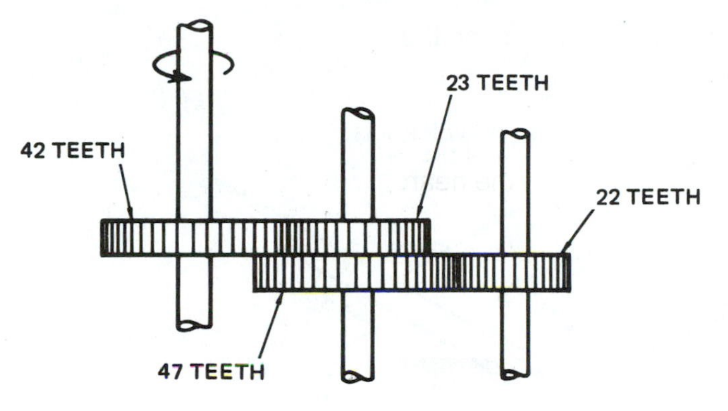

13. The driver gear has 100 teeth and turns clockwise 75 revolutions per minute. Find the speed and the direction of each of these gears.

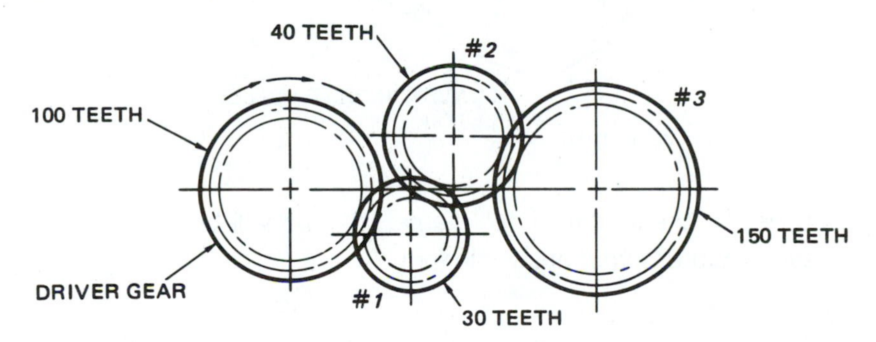

a. Gear #*1* — a. __________

b. Gear #2 — b. __________

c. Gear #*3* — c. __________

CRITICAL THINKING PROBLEMS

1. A certain hydraulic cylinder will extend at 350 inches per minute (ipm) when the input is 76 gallons per minute (gpm). What will be the speed of extension when the input is 118 gpm? ________

2. An overhead conveyor system is configured as shown below. Determine the conveyor speed in feet per minute (fpm). ________

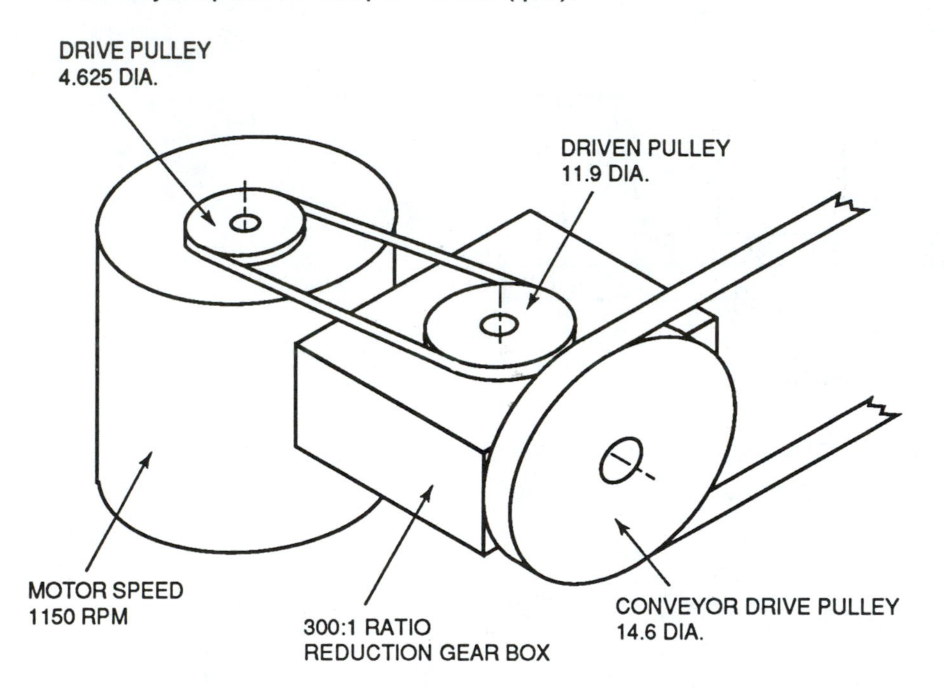

3. If the 11.9" diameter pulley is changed to a 16" diameter, what will be the speed of the conveyor? Express your answer in fpm. ________

Shop Formulas

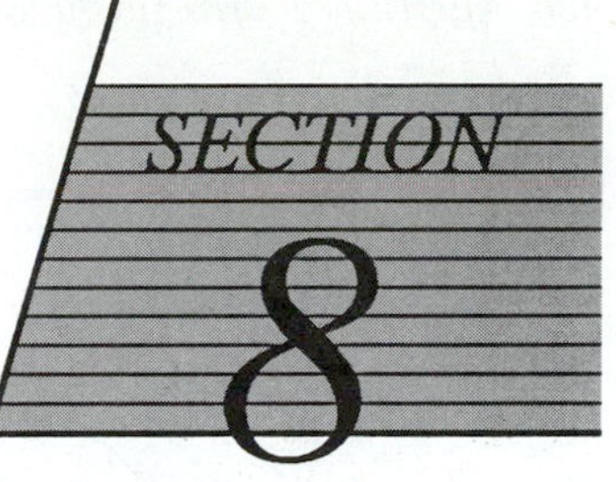

Unit 33 THREAD CALCULATIONS

BASIC PRINCIPLES

- Study and apply these principles of *thread calculations* to these problems.

BRITISH STANDARD

This form is used to designate screw threads belonging to the British Standard series.

[METRIC THREAD DESIGNATION] [TOLERANCE CLASS (PITCH DIAMETER)] [TOLERANCE CLASS (CREST DIAMETER)]

M DIAMETER X PITCH – GRADE POSITION (ALLOWANCE) GRADE POSITION (ALLOWANCE)

SAMPLES: M12 X 1.75 - 5H6H (internal thread)

M8 X 1.25 - 6g (external thread)

Note: If the pitch and crest diameter tolerances are the same, the symbol is used only once.

AMERICAN STANDARD

This form is used to designate screw threads belonging to the American Standard series.

MAJOR DIAMETER – THREADS PER INCH THREAD SERIES – THREAD FIT TYPE OF THREAD

SAMPLES: $\frac{3}{8}$" - 16 UNC - 2A (external thread)

$\frac{7}{16}$" - 20 UNF - 1B (internal thread)

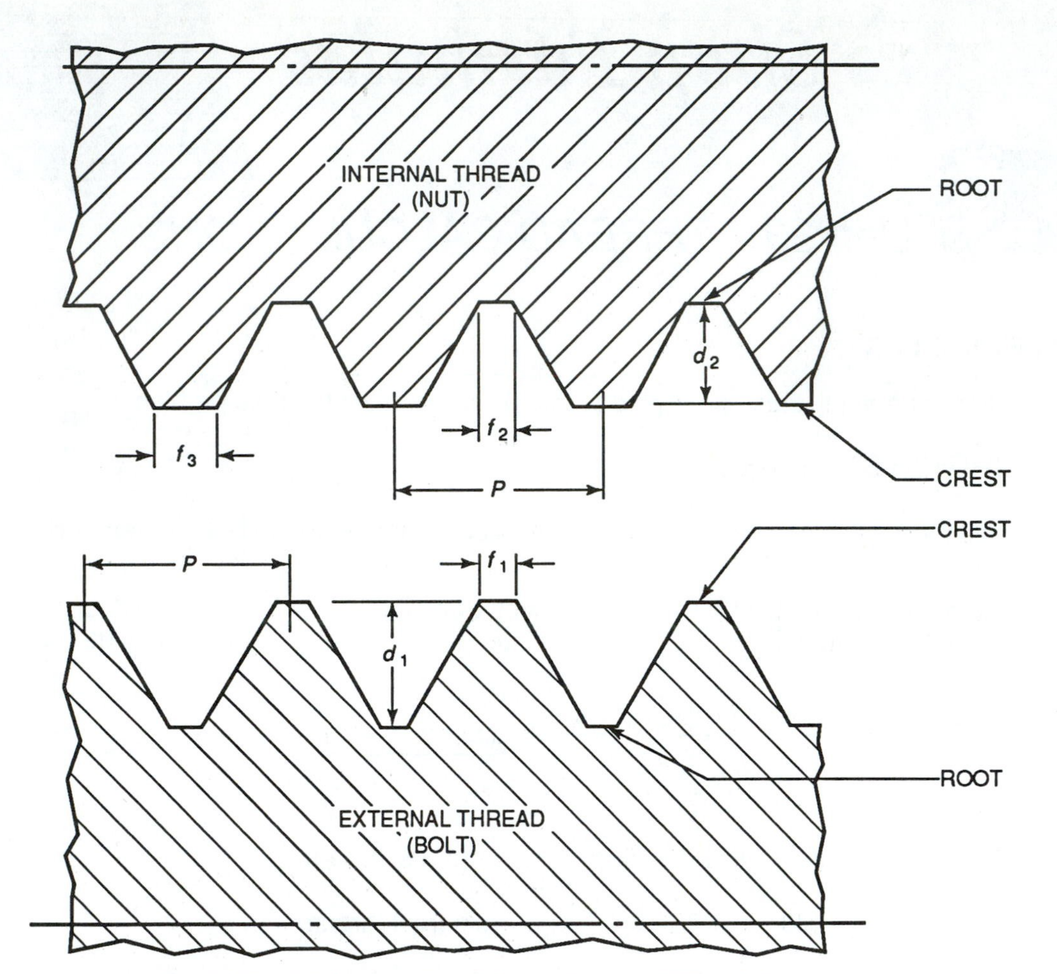

SYMBOL	MEANING
p	pitch
n	number of threads per inch or millimeter
f_1	flat at crest, external thread
f_2	flat at root, internal thread
f_3	flat at crest, internal thread
d_1	depth, external thread
d_2	depth, internal thread

TO FIND	UNIFIED	METRIC
Pitch	$\boldsymbol{p} = \frac{1}{n}$	$\boldsymbol{p} = \frac{1}{n}$
Number of threads per inch or millimeter	$n = \frac{1}{\boldsymbol{p}}$	$n = \frac{1}{\boldsymbol{p}}$
Flat at crest, external thread	$f_1 = 0.125 \times \boldsymbol{p}$	$f_1 = 0.125 \times \boldsymbol{p}$
Flat at root, internal thread	$f_2 = 0.125 \times \boldsymbol{p}$	$f_2 = 0.125 \times \boldsymbol{p}$
Flat at crest, internal thread	$f_3 = 0.250 \times \boldsymbol{p}$	$f_3 = 0.250 \times \boldsymbol{p}$
Depth, external thread	$d_1 = 0.61343 \times \boldsymbol{p}$	$d_1 = 0.61343 \times \boldsymbol{p}$
Depth, internal thread	$d_2 = 0.54127 \times \boldsymbol{p}$	$d_2 = 0.54127 \times \boldsymbol{p}$

When cutting threads it is always good practice to rely on handbook data for the proper pitch diameter for each class of thread rather than these general values.

- Another thread form commonly found in industry is the ACME thread. The Acme thread is generally found where linear motion is required such as lead screws in machine tools. The three classes of American Standard ACME threads are 2G, 3G, and 4G. the Class 2G is the one most commonly used. Refer to standard handbooks for further details of the ACME thread and its uses.

The following figure provides basic information about the ACME thread form.

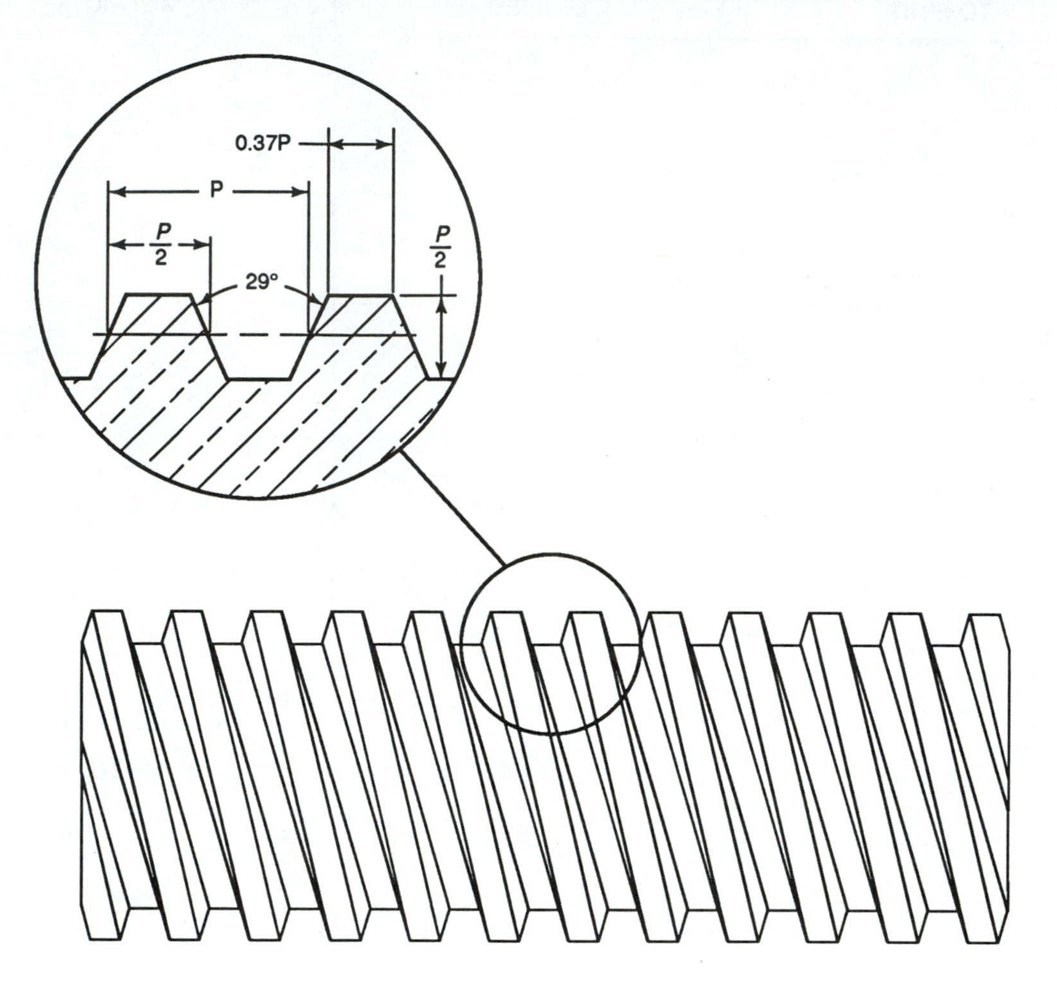

PRACTICAL PROBLEMS

1. Find the values to the nearer thousandth.

	p	$f_{1,2}$	f_3	d_1	d_2
5/8-11					
M10 X 1.5					
1/2-20					
M6 X 1					
5/16-24					
M18 X 2.5					
M12 X 1.75					
1-12					

Note: Use this information for problem 2.

The two basic methods of measuring the pitch diameter of threads are direct and computed. Direct measurement of the pitch diameter uses a screw thread micrometer. When a metric thread is measured with an *English Screw Thread Micrometer*, equivalent measures are used. The equivalent measures are found by using these formulas.

$$\textit{English micrometer reading} = \textit{metric pitch diameter} \times 0.039{,}37$$

or

$$\textit{metric pitch diameter} = \frac{\textit{English micrometer reading}}{0.039{,}37}$$

The number of threads per inch is found by using this formula:

$$\textit{number of threads per inch} = \frac{1}{\textit{metric pitch} \times 0.039{,}37}$$

2. Find the values in the chart to the nearer thousandth.

	SIZE	DESIRED PITCH DIAMETER	NUMBER OF THREADS PER INCH	ENGLISH MICROMETER READING
a.	M48 X 5	44.674 mm		
b.	M10 X 1.5	8.997 mm		
c.	M20 X 2.5	18.307 mm		
d.	M39 X 4	36.271 mm		
e.	M1 X 0.25	0.822 mm		
f.	M27 X 3	25.007 mm		
g.	M2.2 X 0.4	1.732 mm		
h.	M3.5 X 0.6	3.088 mm		

Note: Use this information for problems 3 and 4.

Indirect, or computed, measurements of pitch diameters use these terms, symbols and formulas.

SYMBOL	MEANING
M	measurement over wires
D	major (outside) diameter
p	pitch
w	wire size

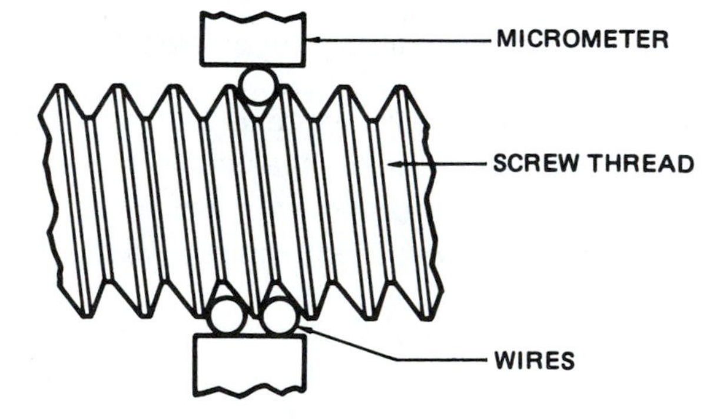

$M = [D - (1.5155 \times \mathbf{p})] + 3w$

The formulas for finding *w* give a best wire size, a minimum wire size, and a maximum wire size.

$$w = 0.57735 \times \mathbf{p} \quad \text{(best)}$$

$$w = 0.5600 \times \mathbf{p} \quad \text{(minimum)}$$

$$w = 0.9000 \times \mathbf{p} \quad \text{(maximum)}$$

Wire size is first calculated using the best wire size. If this size is not available, any size between the minimum wire size and maximum wire size will work.

3. Find the wire sizes to the nearer ten-thousandth.

		BEST	MAXIMUM	MINIMUM
a.	1/2 - 13			
b.	M12 X 1.75			
c.	3/4 - 16			
d.	M6 X 1			
e.	M2 X 2.4			

4. Find the pitch and the best wire size to the nearer ten-thousandth. Using those values, find the measurement over the wires to the nearer ten-thousandth.

		PITCH	BEST WIRE SIZE	MEASUREMENT OVER THE WIRES
a.	1/4 - 28			
b.	M3.5 X 0.6			
c.	5/8 - 11			
d.	9/16 - 12			
e.	M27 X 3			

Note: Use this information for problems 5–14.

The gauge number of *American Standard* machine screws may be expressed as inch diameters. Also, given the inch diameter, the gauge number can be found. The formulas that are used are:

$$\textit{diameter} = (\textit{gauge number} \times 0.013 \text{ in.}) + 0.060 \text{ in.}$$

and

$$\textit{gauge number} = \frac{\textit{diameter} - 0.060 \text{ in.}}{0.013 \text{ in.}}$$

When the gauge is *00*, *000*, or *0000*, the diameter is less than 0.060 inch. The diameter is found by using the formula:

diameter = 0.060 in. - [(*number of zeros in gauge number less 1*) x 0.013 in.]

The number of zeros in the gauge number is:

$$\text{number of zeros in gauge number} = \frac{0.060 \text{ in.} - \text{diameter}}{0.013} + 1$$

5. Find the diameter of a *#6* gauge. ____________
6. Find the gauge for a 0.164-inch diameter. ____________
7. What is the gauge number for 0.125-inch diameter? ____________
8. What is the diameter of a *#3* gauge? ____________
9. Find the diameter for a *#00* gauge. ____________
10. A 0.021-inch diameter means what number gauge? ____________
11. A number 12 gauge means what size diameter? ____________
12. Find the gauge for a 0.073-inch diameter. ____________
13. Find the diameter for a *#2* gauge. ____________
14. What is the diameter of a *#10* gauge. ____________
15. Using the ACME thread figure in the "Basic Principles" section, find the following:
    a. If you have an ACME thread with a 1″ diameter and five threads per inch, what is the pitch (*p*)? ____________
    b. The depth of the thread? ____________
    c. The flat at the top of the thread? ____________

Unit 34 GEAR COMPUTATIONS

PRINCIPLES

- Study and apply these principles of *gear calculations* to these problems.

  Gears are used to transmit rotary motion and power from one shaft to another shaft. Gears are designed in standard types and sizes and contain standard parts.

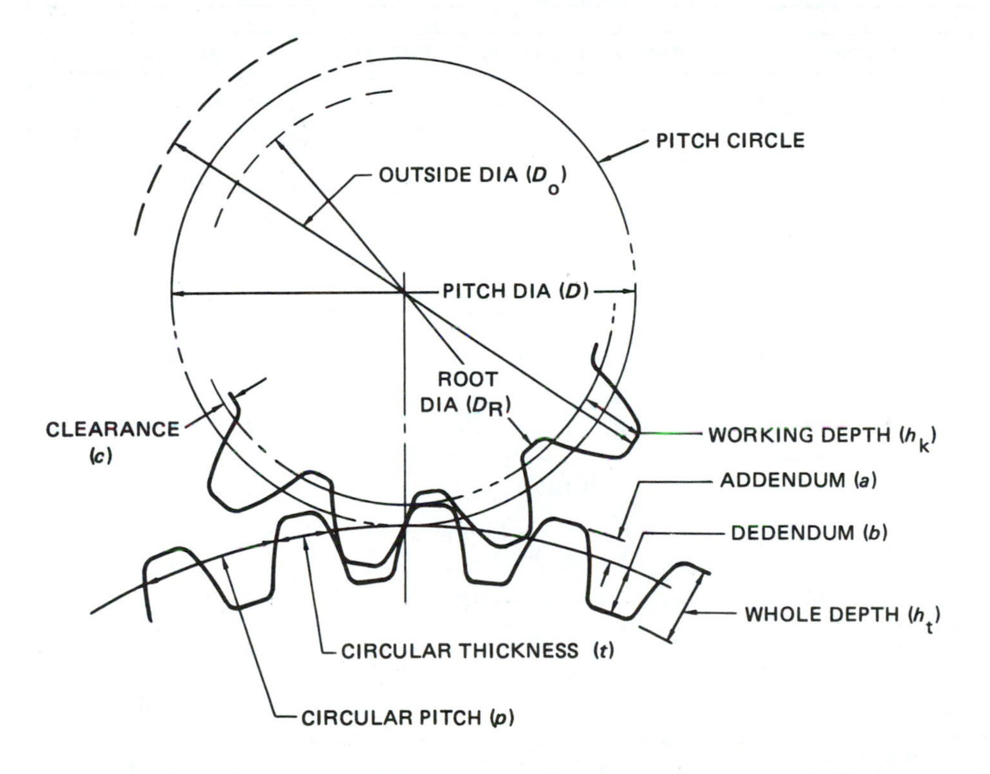

The *American Standard* gears use the diametrical pitch or the circular pitch for spur gears computations. The *British Standard* gears use the module. The *module* is the number of millimeters of pitch diameter for each tooth in the gear. This means that the module actually defines the size of the gear tooth.

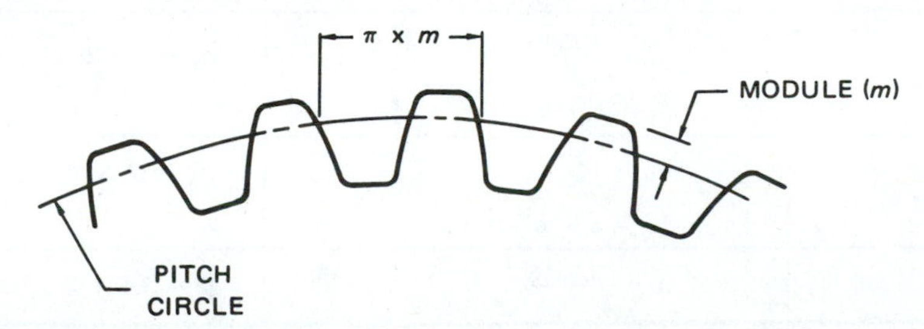

SYMBOL	MEANING
D	pitch diameter
D_o	outside diameter
D_R	root diameter
P	diametral pitch
p	circular pitch
t	circular thickness

SYMBOL	MEANING
a	addendum
b	dedendum
c	clearance
h_t	whole depth
h_k	working depth
n	number of teeth
m	module

TO FIND	AMERICAN NATIONAL STANDARD	METRIC
Module	....	$m = \frac{D}{n}$
Diametral pitch	$P = \frac{n}{D}$	$p = \frac{1}{m}$
Pitch diameter	$D = \frac{n}{P}$ $D = \frac{D_o \times n}{(n + 2)}$	$D = n \times m$
Number of teeth (expressed as a whole number)	$n = P \times D$	$n = \frac{D}{m}$
Addendum	$a = \frac{1}{P}$	$a = m$
Dedendum (preferred)	$b = \frac{1.250}{P}$	$b = 1.250 \times m$
Clearance (preferred)	$c = \frac{0.250}{P}$	$c = 0.250 \times m$
Clearance (minimum)	$c = \frac{0.157}{P}$	$c = 0.157 \times m$
Circular thickness-Basic	$t = \frac{1.5708}{P}$	$t = 1.570\ 8 \times m$
Root diameter	$D_R = \frac{(n - 2.5)}{P}$ $D_R = D - (2 \times b)$	$D_R = D - (2.5 \times m)$
Outside diameter	$D_o = \frac{(n + 2)}{P}$ $D_o = D + (2 \times a)$	$D_o = m \times (n + 2)$ $D_o = D + (2 \times m)$
Whole depth (preferred)	$h_t = \frac{2.250}{P}$ $h_t = a + b$	$h_t = a + b$
Working depth	$h_k = \frac{2}{P}$ $h_k = a + b - c$	$h_k = 2 \times a$
Circular pitch	$p = \frac{3.1416}{P}$	$p = 3.141\ 6 \times m$

PRACTICAL PROBLEMS

1. A manufacturer must make a gear with 48 teeth and an outside diameter of $6\frac{1}{4}$ inches. Find the values on this chart. Round to the nearer thousandth, when necessary.

	DIMENSION	VALUE
a.	Pitch diameter (D)	
b.	Diametral pitch (P)	
c.	Addendum (a)	
d.	Dedendum (b)	
e.	Clearance (c)	
f.	Circular thickness (t)	
g.	Circular pitch (p)	
h.	Whole depth (h_t)	
i.	Working depth (h_k)	
j.	Root diameter (D_R)	

2. A machinist receives a broken gear that has 7 teeth in a 45° segment. The machinist calculates the pitch diameter to be 3.5 centimeters (350 millimeters). Find the other calculations that the machinist needs. Express the answers to four decimal places, when necessary.

	DIMENSION	VALUE
a.	m	
b.	a	
c.	b	
d.	c	
e.	t	
f.	D_o	
g.	D_R	
h.	h_t	
i.	h_k	
j.	p	

3. A gear has a diametral pitch of 8 and a pitch diameter of 5.750 inches. Find, to the nearer thousandth, the values needed to make this gear.

	DIMENSION	VALUE
a.	n	
b.	a	
c.	b	
d.	c	
e.	t	
f.	D_O	
g.	D_R	
h.	h_t	
i.	h_k	
j.	p	

4. A manufacturer must remake a *German* gear with 18 teeth and an outside diameter of 320 mm.

   a. What is the module? a. ___________

   b. Find the pitch diameter. b. ___________

   c. What is the circular pitch? c. ___________

5. A gear must have 44 teeth and a circular pitch of 0.2165 inch.

   a. Find the circular pitch to the nearer ten-thousandth millimeter. a. ___________

   b. What is the metric module to the nearer thousandth millimeter? c. ___________

   c. Find the pitch diameter. c. ___________

   d. What is the tooth thickness to the nearer thousandth millimeter? d. ___________

6. A 60° segment of a broken gear has exactly 16 teeth. The metric module is 1.25 mm.

   a. What is the pitch diameter? a. ___________

   b. What is the outside diameter? b. ___________

   c. Find, to the nearer thousandth millimeter, the whole depth. c. ___________

Unit 35 SPEED AND FEED CALCULATIONS FOR CYLINDRICAL TOOLS

BASIC PRINCIPLES

- Study and apply these principles of cutting speed calculations for cylindrical tools to these problems.

  When cylindrical cutting tools — such as taps, reamers, counterboring tools, and milling cutters — are used in the machine shop, it is necessary to know the cutting speed and the spindle speed.

  The *cutting speed* is the distance that a point on the circumference of the tool will travel in one minute. The cutting speed is expressed in *feet per minute* (*fpm*) or *meters per minute* (*m/min*). The *spindle speed* is the number of times that the cylindrical tool will turn in one minute. It is expressed in *revolutions per minute*.

SYMBOL	MEANING	AMERICAN STANDARD UNITS	METRIC UNITS
V	cutting speed	feet per minute (fpm)	meters per minute (m/min)
D	diameter	inches (in.)	millimeters (mm)
N	spindle speed	revolutions per minute (rpm)	revolutions per minute (r/min)

TO FIND	AMERICAN STANDARD UNITS	METRIC UNITS
N	$N = \dfrac{12 \times V}{3.1416 \times D}$	$N = \dfrac{1{,}000 \times V}{3.1416 \times D}$
V	$V = \dfrac{3.1416 \times D \times N}{12}$	$V = \dfrac{3.1416 \times D \times N}{1{,}000}$
D	$D = \dfrac{V \times 12}{3.1416 \times N}$	$D = \dfrac{V \times 1{,}000}{3.1416 \times N}$

The diameter that is selected should be that of the rotation member. For drilling or reaming, use the diameter of the drill or the reamer. For milling, use the diameter of the cutter.

The formulas for cutting speed and spindle speed may also be used for turning or boring on a lathe. In these operations the workpiece rotates and, therefore, the diameter of the workpiece (or the diameter of the hole) is used. Cutting speed for turning or boring then refers to the rate at which the tool removes the stock from the surface, or the distance that a point on the workpiece travels in one minute.

PRACTICAL PROBLEMS

1. Determine the rpm required to machine a 310-mm-diameter gear blank if the cutting speed is 22 meters per minute. Round the answer to the nearer whole number. __________

2. Calculate the rpm required when turning a 356-mm-diameter pulley with a cutting speed of 190 meters per minute. Round answer to nearer whole number. __________

3. Find the rpm required to obtain a cutting speed of 18 meters per minute for the cylinder shown. Round the answer to the nearer whole number. __________

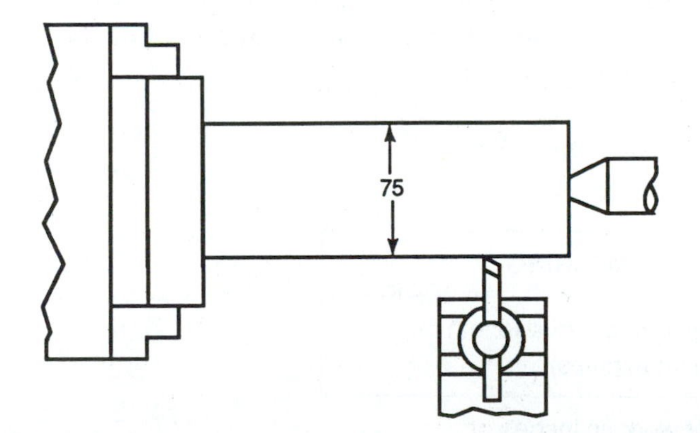

4. Determine the diameter of the stock being turned if the cut meter shows the surface speed to be 65 fpm and the stock is being turned at 65 rpm. Round answer to the nearest ten-thousandth of an inch. __________

5. Find the diameter of the part being machined if the cutting speed is 45 meters per minute and the rpm is 100. Round answer to the nearer hundredth millimeter. __________

6. Determine the rpm required when turning a 97-mm-diameter shaft at 25 meters per minute. Round answer to nearest whole number. __________

7. This cast iron cylinder is to be turned at 50 feet per minute. The dimensions of the cylinder are in inches. Express each answer to the nearer whole number.

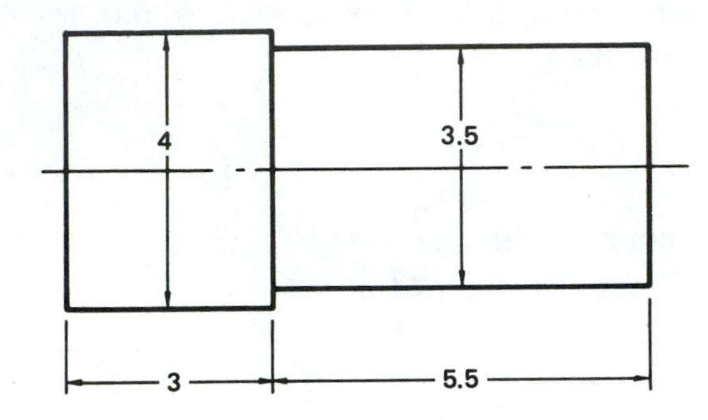

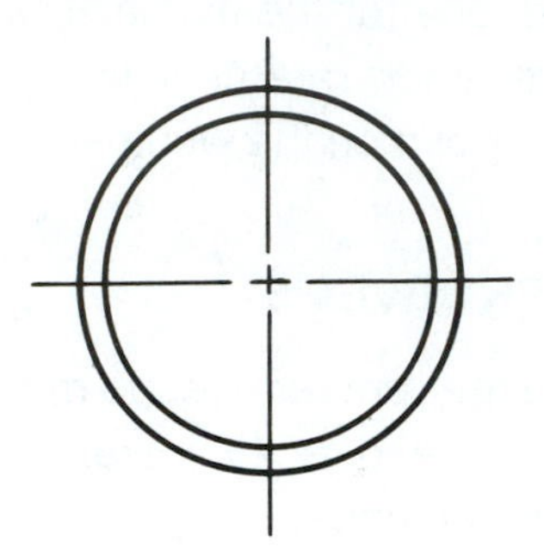

   a. Find the rpm required for the larger step. a. __________

   b. Find the rpm required for the smaller step. b. __________

8. Determine the rpm required for a $4\frac{1}{4}''$-inch shell end mill when milling the edge of a 1-inch-thick carbon steel plate with a cutting speed of 90 fpm. Round answer to nearest whole number. __________

Note: Use this information for problems 9–12.

SYMBOL	MEANING
T	time required to turn a length of material (in minutes)
L	length of work (in inches or millimeters)
N	spindle speed (in revolutions per minute)
f	feed (in inches or millimeters)

$$T = \frac{L}{N \times f}$$

9. A 10-inch-diameter cast iron pulley is shown.

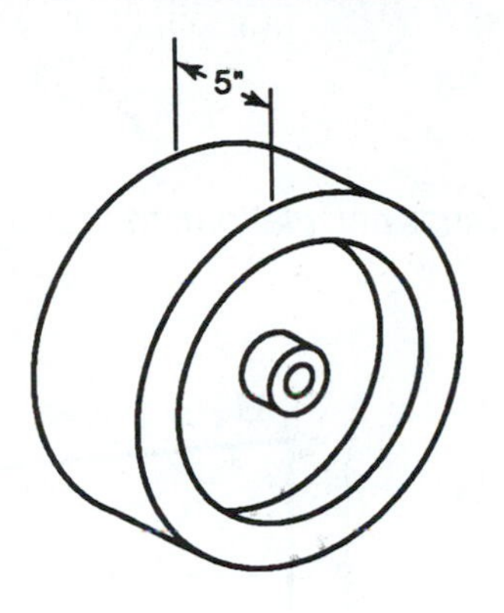

a. If the cutting speed is 120 fpm, find the spindle speed to the nearer whole number. a. __________

b. With a feed of 0.018 inch, how long will it take to make one cut across the face of the pulley? Round the answer to the nearer hundredth minute. b. __________

10. Find, to the nearer hundredth minute, the time required for drilling a $\frac{3}{8}$-inch-diameter hole through a piece of $\frac{1}{2}$-inch cold rolled steel. A high-speed steel drill is used with a cutting speed of 70 feet per minute and a feed of 0.005 inch. __________

11. With a 0.003-inch feed, how long does it take to drill a $\frac{3}{16}$-inch hole through a $1\frac{3}{4}$-inch piece of carbon steel? The cutting speed of the drill is 90 feet per minute. Round the answer to the nearer hundredth minute. __________

12. A tool steel machine spindle is $1\frac{1}{4}$ inches in diameter, is 17 inches long, and has a cutting speed of 40 feet per minute. Find, to the nearer second, the time required to take one roughing cut. The feed is 0.020 inch. Allow 50 seconds for changing the lathe dog. __________

Unit 36 TAPER CALCULATIONS

BASIC PRINCIPLES

- Study and apply these principles of *taper calculations* to these problems.

SYMBOL	MEANING
tpi	taper per inch
tpf	taper per foot
D	diameter, larger end
d	diameter, small end
L	length (in inches)

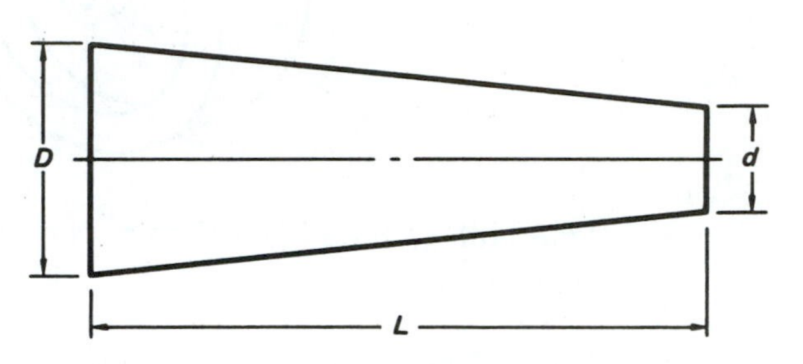

TO FIND	KNOWN	FORMULA
tpf	*tpi*	$tpf = tpi \times 12$
	D, d, L	$tpf = 12\left(\frac{D - d}{L}\right)$
tpi	*tpf*	$tpi = \frac{tpf}{12}$
	D, d, L	$tpi = \frac{D - d}{L}$
d	*D, L, tpf*	$d = D - \left[L\left(\frac{tpf}{12}\right)\right]$
D	*d, L, tpf*	$D = d + \left[L\left(\frac{tpf}{12}\right)\right]$
L	*D, d, tpf*	$L = 12\left(\frac{D - d}{tpf}\right)$ or $L = \left(\frac{D - d}{tpi}\right)$

PRACTICAL PROBLEMS

1. The piece shown in the drawing has a taper of 1.105 inches per foot. Find dimension ***d*** to the nearer thousandth inch. __________

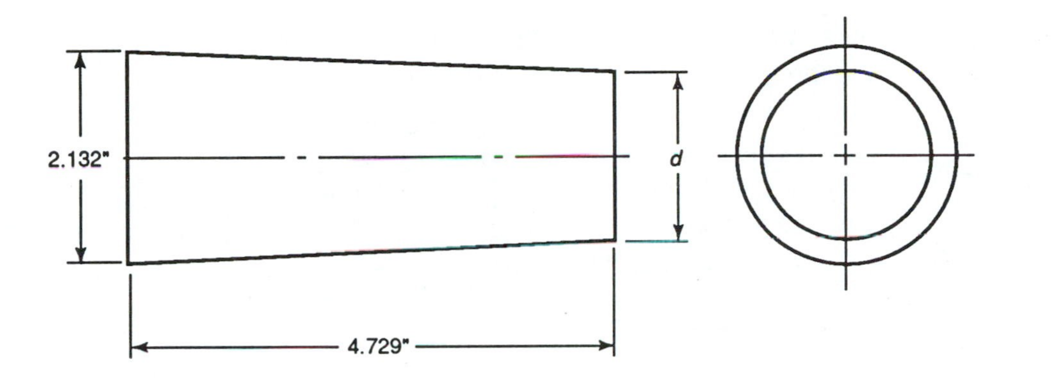

2. Find the length of a taper that is 0.486 inch larger at one end than the other. It tapers at the rate of 0.750 inch per foot. __________

3. If a piece of work tapers at the rate of $\frac{5}{8}$ inch to the foot, how much does it taper in $19\frac{3}{4}$ inches? Round the answer to four decimal places. __________

4. Use this diagram to find the values in the chart. Round each answer to four decimal places when necessary. __________

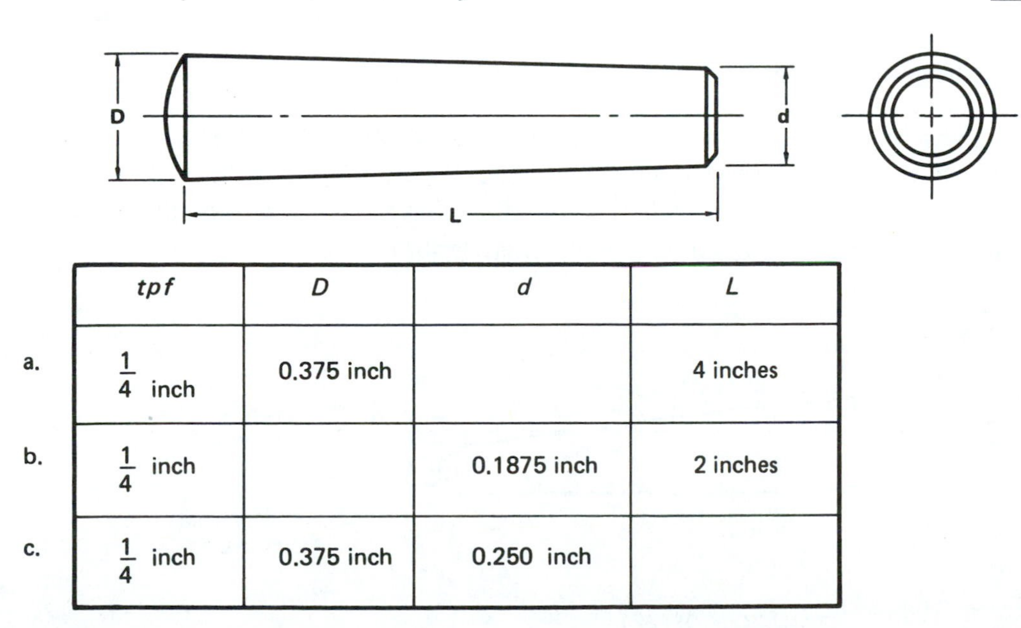

	tpf	*D*	*d*	*L*
a.	$\frac{1}{4}$ inch	0.375 inch		4 inches
b.	$\frac{1}{4}$ inch		0.1875 inch	2 inches
c.	$\frac{1}{4}$ inch	0.375 inch	0.250 inch	

5. A piece $11\frac{1}{2}$ inches long is turned with a taper of 0.602 inch per foot. The length of the taper is $\frac{3}{4}$ the total length of the piece. The diameter at the large end is $7\frac{1}{16}$ inches. What is the small diameter to the nearer thousandth inch? __________

6. The small end of *Number 11 Brown & Sharpe* taper is $1\frac{1}{4}$ inches. The length is $6\frac{3}{4}$ inches and the tpf is approximately 0.500 inch.

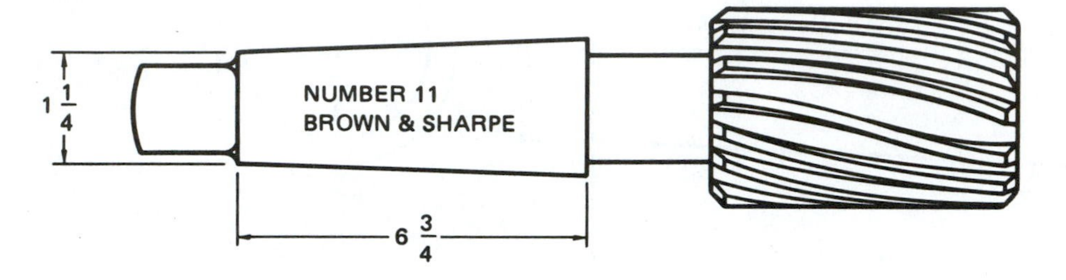

a. Determine the amount of taper per inch. Round the answer to four decimal places. a. __________

b. Determine the diameter of the large end to the nearer thousandth inch. b. __________

7. At one end a workpiece is 0.725 inch. At the other end it measures 0.580 inch. If the taper is $\frac{1}{4}$ inch per foot, what is the length of the piece? __________

Note: Use this information for problems 8–11.

The machine operator often makes calculations for offsetting the tailstock to turn a taper that extends only part of the length of the piece. When doing this, the machine operator must bear in mind that the tailstock must be offset just as much as though the taper were to continue at the same rate throughout the entire length.

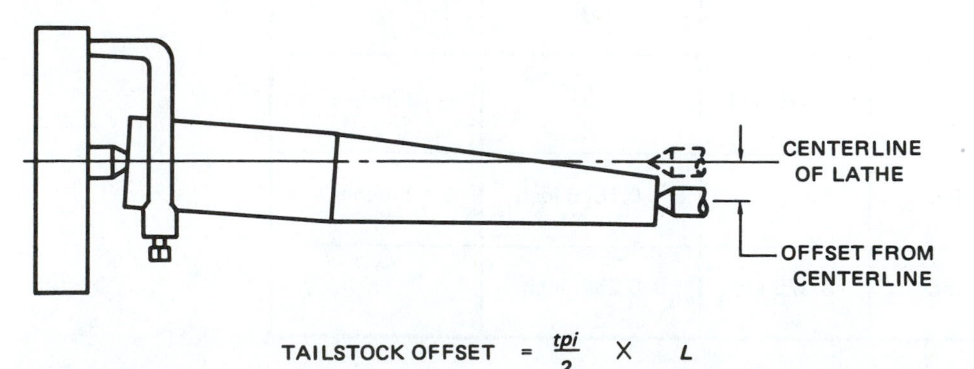

$$\text{TAILSTOCK OFFSET} = \frac{tpi}{2} \times L$$

8. A piece of work, 12 inches long, has a taper of $\frac{7}{16}$ inch per foot. How much should the tailstock be offset to turn the taper? Round the answer to four decimal places. __________

9. A taper pin is 0.591 inches in diameter at its large end and is $5\frac{1}{4}$ inches long. The rate of taper is $\frac{1}{4}$ inch per foot.

   a. What is the diameter at the small end to the nearer thousandth? a. __________

   b. How much should the tailstock be set to turn the taper? Express the answer to the nearer ten-thousandth inch. b. __________

10. A *Brown & Sharpe* taper is 5 inches long. The taper per foot is $\frac{1}{2}$ inch. Determine, to the nearer ten-thousandth inch, the tailstock offset. __________

11. Determine the tailstock offset needed to turn a *Number 3 Brown & Sharpe* taper on a 14-inch-long milling machine arbor. The taper per foot is $\frac{1}{2}$ inch. Express the answer to four decimal places. __________

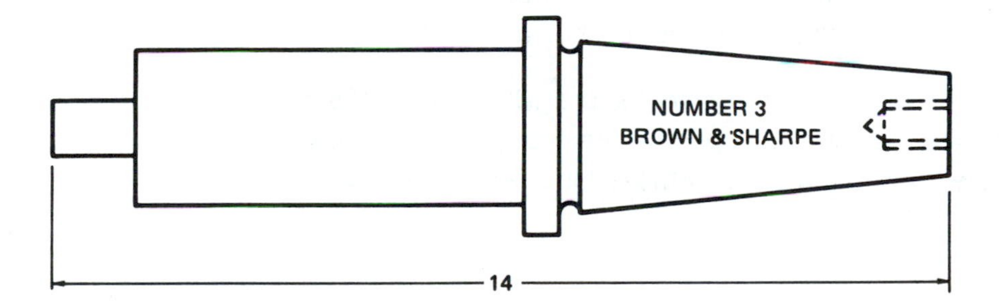

CRITICAL THINKING PROBLEMS

1. An engineer is designing a machine that requires a lead screw that will move a dovetail slide 1″ in 20 turns. How many threads per inch (tpi) are required on the lead screw? What common diameter and tpi might be used? __________

2. Determine the rpm required to turn the step on the steel shaft shown if the cylinder is turned at 100 feet per minute (fpm). __________

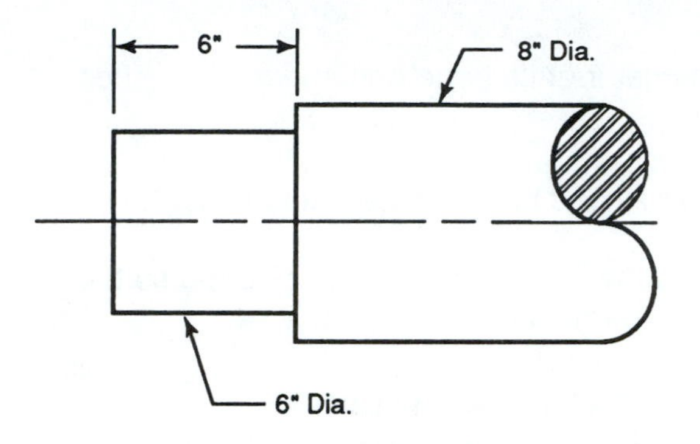

3. Assuming a feed of 0.040″ per revolution and 10 passes, how long will it take to machine the step on the shaft in question #2? __________

4. If the shaft in question #2 is 3′ long and the cost of the shaft is 86 cents per pound, how much will the material for each shaft cost? __________

5. Determine how much it will cost to manufacture 100 of the shafts shown in question #2 if the machine and operator are charged out at $25.00 per hour and it takes 2.5 minutes to remove and replace a part in the machine. __________

Powers, Roots, and Equations

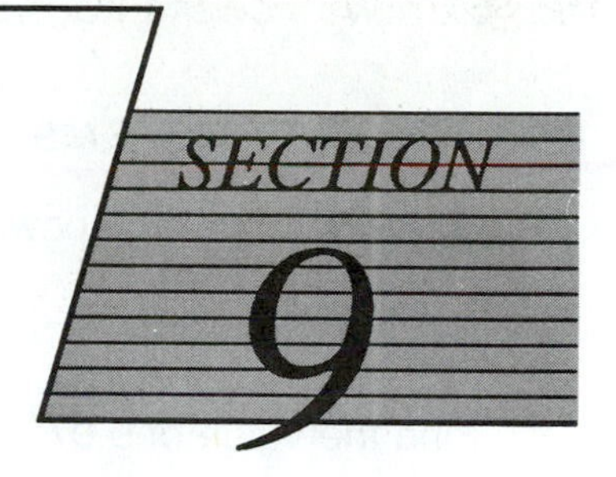

Unit 37 POWERS

BASIC PRINCIPLES

- *Powers* or *exponents* are a simple way of indicating how many times a number is to be multiplied by itself.

  Example:

  $$4^5 = 4 \times 4 \times 4 \times 4 \times 4 = 1{,}024$$

- When a number is multiplied by itself, the number is said to be *squared*. When using this for area measurement, the result will be expressed in square units.

  Example:

  $$(5 \text{ inches})^2 = 5'' \times 5'' = 25 \text{ square inches} = 25 \text{ in}^2$$

  When a number is multiplied by itself three times, the number is said to be *cubed*. When using this for volume measurement, the result is expressed in cubic units.

  Example:

  $$(4 \text{ feet})^3 = 4' \times 4' \times 4' = 64 \text{ cubic feet} = 64 \text{ ft}^3$$

- When multiplying terms containing exponents, the exponent of the multiplicand and the exponent of the multiplier are added. The resulting sum is used as the exponent in the answer.

  Example:

  $$4^2 \times 4^3 = 4^5 = 1{,}024$$

- When dividing terms containing exponents, the exponent of the divisor is subtracted from the exponent of the dividend. The resulting exponent is used as the exponent of the quotient.

  Example:

  $$\frac{5^5}{5^3} = 5^2 = 25$$

PRACTICAL PROBLEMS

1. Raise 5 to the third power. ____________

2. What is the cube of 15.3? ____________

3. Find the value of 0.07^2. ____________

4. Find 3.1^4. ____________

5. Find 1.07^5. ____________

6. The fraction ¾ raised to the second power is written $(3/4)^2$

   What is the value of $(3/4)^2$? ____________

Note: Use this information for problems 7–10.

SYMBOL	MEANING
A	area
V	volume
r	radius
D	diameter

The area of a circle is:

$A = 3.1416 \times r^2$ or $A = 0.7854\ D^2$

The volume of a sphere is:

$$V = \frac{3.1416 \times 4 \times r^3}{3} \text{ or } V = \frac{3.1416 \times D^3}{6}$$

$$r = \frac{D}{2} \text{ and } D = 2r$$

7. Find the area of a circle that is 6.5 cm in diameter. ____________

8. What is the volume of a sphere that is 6.5 cm in diameter? ____________

9. Find the area of a circle with a radius of 7.94 mm. Round the answer to four decimal places. ____________

10. What is the volume of a sphere with a diameter of 7.94 mm? Round the answer to four decimal places. ____________

11. Find the value for each expression. Express each answer in exponential form.

    a. $3^3 \times 3^6$ a. ____________

    b. $N^2 \times N \times N^3$ b. ____________

c. $p^4 \times p^3 \times p^2$ c.__________

d. $19^2 \times 19^2 \times 19^3$ d.__________

12. Find the value of each expression. Express each answer in exponential form.

a. $2^3 \div 2^2$ a.__________

b. $B^9 \div B^6$ b.__________

c. $\frac{x^3}{x^2}$ c.__________

d. $\frac{a^7}{a^3}$ d.__________

13. Find the value of each expression. Express each answer in exponential form.

a. $\frac{a^5 \times a^2}{a}$ a.__________

b. $\frac{7^2 \times 7^3 \times 7^5}{7 \times 7^2}$ b.__________

c. $\frac{B^5 \times B}{B^3 \times B^2}$ c.__________

d. $\frac{5^2 \times 5^3 \times 5^4}{5^2}$ d.__________

14. Find the value of each expression.

a. 5^2 a.__________

b. 4^2 b.__________

c. $7^4 \times 7^2$ c.__________

d. 9^3 d.__________

e. $2^2 \times 2^2$ e.__________

f. $\frac{3^5}{3 \times 3^2}$ f.__________

g. 4.79^2 g. ________

h. 0.092^3 h. ________

i. 7^4 i. ________

j. $4^3 \div 4^2$ j. ________

k. $\frac{2^3 \times 2^2}{2}$ k. ________

Unit 38 SQUARE ROOTS

BASIC PRINCIPLES

- A *square root* is a number that when squared will produce a given number. For example, 3 is the square root of 9. The square root of a number can be determined by several methods including the use of *Tables of Square Roots* and more often, the scientific calculator. Scientific calculators generally have a square root key on them. To determine the square root of a number, simply key the number into the calculator and press the square root ($\sqrt{x}$) key. The square root is then displayed on the readout.

REVIEW PROBLEMS

Use the Table of Square Roots in the appendix to find the square roots of the following numbers:

a. $\sqrt{4}$ = b. $\sqrt{33}$ = c. $\sqrt{78}$ =

Use the calculator to determine the square roots of the following numbers:

d. $\sqrt{34.78}$ = e. $\sqrt{125}$ = f. $\sqrt{1{,}024}$ =

PRACTICAL PROBLEMS

1. Find each square root.

   a. $\sqrt{4}$ a. ____________

   b. $\sqrt{25}$ b. ____________

   c. $\sqrt{64}$ c. ____________

2. Find the value of each expression.

   a. $\sqrt{144}$ a. ____________

   b. $\sqrt{225}$ b. ____________

   c. $\sqrt{625}$ c. ____________

   d. $\sqrt{53{,}361}$ d. ____________

   e. $\sqrt{516{,}961}$ e. ____________

3. Find the square root of 1,234,321. ________

Note: Use this information for problems 4–7.

To find the square root of a common fraction when the numerator and denominator are perfect squares, find the square root of the numerator and the square root of the denominator.

$$\sqrt{\frac{4}{25}} = \frac{\sqrt{4}}{\sqrt{25}} = \frac{2}{5}$$

4. What is the value of $\sqrt{49/169}$? ________

5. Find the square root of $4/9$. ________

6. Find the square root of $64/81$. ________

7. Find the square root of $9/25$. ________

Note: Use this information for problems 8–12.

The area of a circle is equal to 3.1416 times the square of its radius. It is written $A = 3.1416 \times r^2$.

The radius of a circle: $\sqrt{\frac{A}{3.1416}}$

Round the answer to four decimal places.

8. If a circle has an area of 1,790 mm^2, what is its radius? ________

9. The area of a circle is 28.98 cm^2. Find its diameter. ________

10. What is the diameter of a circle which has an area of 45.97 square inches? ________

11. What is the radius of a circle whose area is 1.356 m^2? ________

12. The area of a circle is 5,789.456,9 mm^2. What is its diameter? ________

Unit 39 EXPRESSIONS AND EQUATIONS

BASIC PRINCIPLES

- When evaluating expressions it must be remembered that operations must be performed in the proper order or the result will not be correct. The "Priority of Operations" is as follows:

  1. Any operation or operations within parentheses or other grouping symbols.
  2. Exponentiation
  3. Multiplication/division (left to right)
  4. Addition/subtraction (left to right)

  Example: Evaluate the following expression:

$$\frac{\frac{(2^2)(3^2)(6)}{(1+7)}}{4}$$

  Step 1: Evaluate all operations inside parentheses:

$$\frac{\frac{(4)(9)(6)}{(8)}}{4}$$

  Step 2: Evaluate all multiplications and divisions:

$$\frac{216}{2} = 108$$

- When solving equations, two points to be remembered are:

  1. The unknown variable must be isolated on one side of the equation for the equation to be considered solved.
  2. The same operation may be performed to both sides of the equation and the equation will remain an equality.

  Example: Solve the following equation for x:

$$x - 5 = 12$$

Add five to both sides of the equation. This operation will effectively cancel the -5 from the left side of the equation, isolating the unknown on the left of the equal sign.

$$x - 5 + 5 = 12 + 5$$
$$x = 17$$

This can be accomplished by transposing the -5 to the right side of the equation and changing its sign.

$$x - 5 = 12$$
$$x = 12 + 5$$
$$x = 17$$

Example: Solve the following equation for x:

$$3x + 6 = 22 - x$$
$$3x + x = 22 - 6$$
$$4x = 16$$

To complete the solution of this problem, divide both sides of the equation by 4.

$$\frac{4x}{4} = \frac{16}{4}$$
$$x = 4$$

If the inside diameter (ID) and wall thickness (T) of a piece of tubing is known, a formula can be used to calculate the outside diameter (OD):

$$OD = ID + 2T$$

where OD = outside diameter

ID = inside diameter

T = wall thickness

Example: Calculate the outside diameter (OD), when ID = 1.5″ and T = 0.0625″.

$$OD = ID + 2T$$

substitute values $OD = 1.5'' + 2(0.0625'')$

$$OD = 1.5'' + 0.125''$$
$$OD = 1.625''$$

If the same formula is used and the OD is known, the ID can be calculated by following the same method.

Example: Calculate the inside diameter (ID″, when OD = 2.250″ and T = 0.250″

	OD	=	ID + 2T
substitute values	2.250″	=	ID + 2(0.250″)
	2.250"	=	ID + 0.5″
isolate unknown variable	-ID	=	-2.250″ + 0.5″
	-ID	=	-1.750″
change signs	-1(-ID)	=	-1(-1.750″)
	ID	=	1.750″

PRACTICAL PROBLEMS

1. $4\,(1 + 2)$ ________

2. $(4 \times 4)\,(1 + 2)$ ________

3. $3\left[\dfrac{(2+4)}{(8-6)}\right]$ ________

4. $3\left\{2\,[\,(1 + 2) + (4 - 2)\,]\right\}$ ________

5. $\dfrac{(2+1+4+7)}{(6-4)}$ ________

6. $3\left\{\dfrac{4[2(8-4)+(2+1)]}{[8(7-6)+3]}\right\}$ ________

7. $\dfrac{5 \cdot 5 \cdot 5}{5 \cdot 5}$ ________

8. $2\left[\left(\dfrac{8 \times 8}{16}\right) + 2\left(\dfrac{72 - 18}{4 - 1}\right)\right]$ ________

9. $\dfrac{(1 + 1 + 4) + (3 + 2 + 1) + (3 \times 5 \times 5)}{3}$ ________

10. $\dfrac{45}{5(9 - 3) + 5(6 - 3)}$ ________

11. Solve for x : $x + 4 = 10$ __________

12. Solve for x : $x - 7 = 8 + 2$ __________

13. Solve for x : $2x = 48$ __________

14. Solve for x : $\frac{2x}{4} = 12$ __________

15. Solve for n: $\frac{n+4}{3} = 7$ __________

16. Solve for x: $9x + 3x = 108$ __________

17. Solve for a: $\frac{3a + 4a + 2a}{2} = 20.25$ __________

CRITICAL THINKING PROBLEMS

1. In the technical world, very large or small numbers are expressed in "scientific notation" or powers of ten.

   Example:

   $$6236 = 6.236 \times 10^3$$

   $$.000036 = 3.6 \times 10^{-5}$$

   If your calculator can only display 10 digits it would be impossible to display or use a number with more than 10 digits on the calculator. Most scientific calculators can be used with scientific notation so that very large or small numbers can be used in calculations.

   Convert the following numbers to scientific notation:

   a. 93,000,000 = __________

   b. 186,000 = __________

   c. .001 = __________

   d. .0000000631 = __________

Using your calculator's scientific notation capability, do the following calculations:

e. 3.14 x .000012 = __________

f. .001 x 6000 = __________

g. .0000000000668 x 43660000000 = __________

2. The Pythagorean theorem states that in a right triangle

$$c = \sqrt{a^2 + b^2}$$

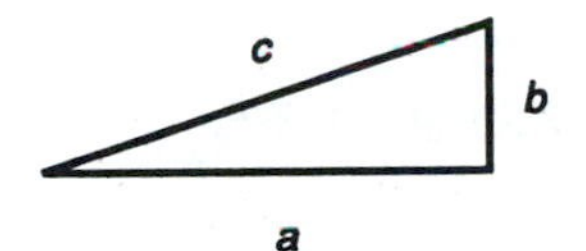

Draw the following triangle to scale and measure side *c*.

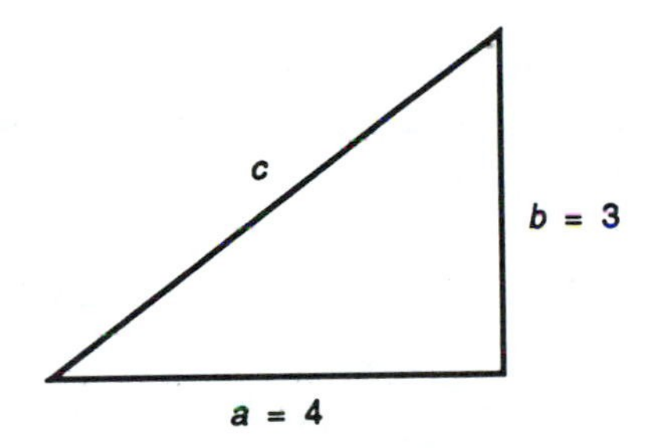

Does the Pythagorean theorem hold true for this triangle?

$$\begin{aligned} c &= \sqrt{a^2 + b^2} \\ &= \sqrt{4^2 + 3^2} \\ &= \sqrt{16 + 9} \\ &= \sqrt{25} \\ c &= 5 \end{aligned}$$

Draw the following triangles, measure side *c,* and verify your answer using the Pythagorean theorem.

a. Side a = 12.69, side b = 3.21. c = ____

b. Side a = 1.44, side b = 16.66, c = ____

c. Side a = 14.12, side b = 15.27, c = ____

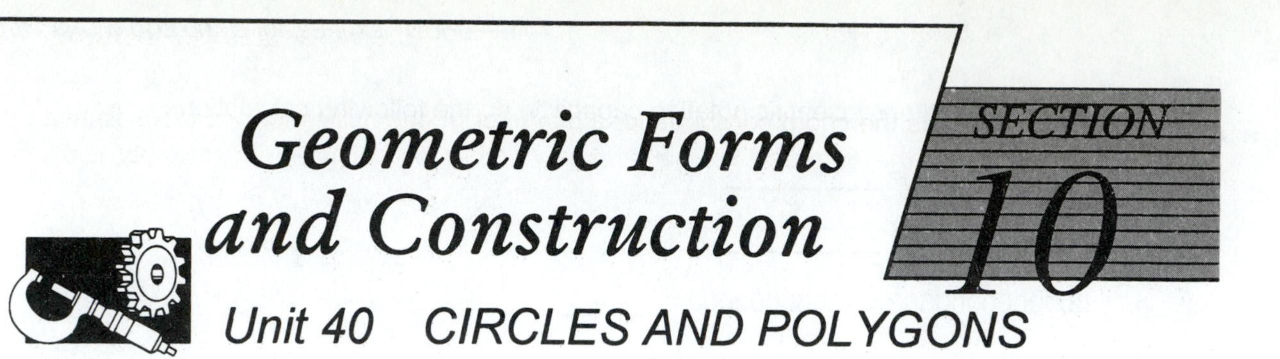

Geometric Forms and Construction

Unit 40 CIRCLES AND POLYGONS

BASIC PRINCIPLES

The machinist must work with circles and polygons on a daily basis. The ability to lay out and fabricate circular parts is a skill that is of value to the machinist. The following problems are typical of those often found in the machine shop.

Equally spacing holes around a bolt circle is a task a machinist must perform often. To make sure the spacing is as accurate as possible, the machinist can use either chords or horizontal and vertical distances to space the holes.

A *chord* is a straight line that connects two points of an arc.

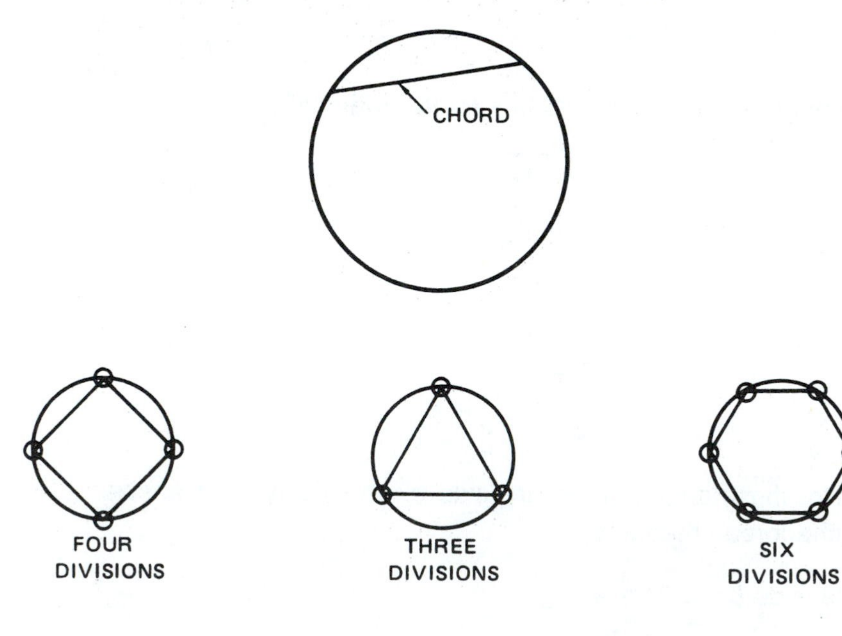

This table lists the chord constants for a diameter of one millimeter/one inch. To use the chart, select the constant for the desired number of divisions and multiply it by the diameter of the bolt circle. This results in the length of the chord to be used.

DIVISIONS	DEGREE	CONSTANT*
3	120°	0.8660
4	90°	0.7071
5	72°	0.5878
6	60°	0.5000
8	45°	0.3827
10	36°	0.3090
12	30°	0.2588
15	24°	0.2079
18	20°	0.1736
20	18°	0.1564
24	15°	0.1305
30	12°	0.1045
32	$11\frac{1}{4}°$	0.0980
36	10°	0.0872

*Based on a hole diameter of 1 mm/1 in.

PRACTICAL PROBLEMS

1. This cover plate is to be manufactured. What is the length of the chord needed for these holes in millimeters? ____________

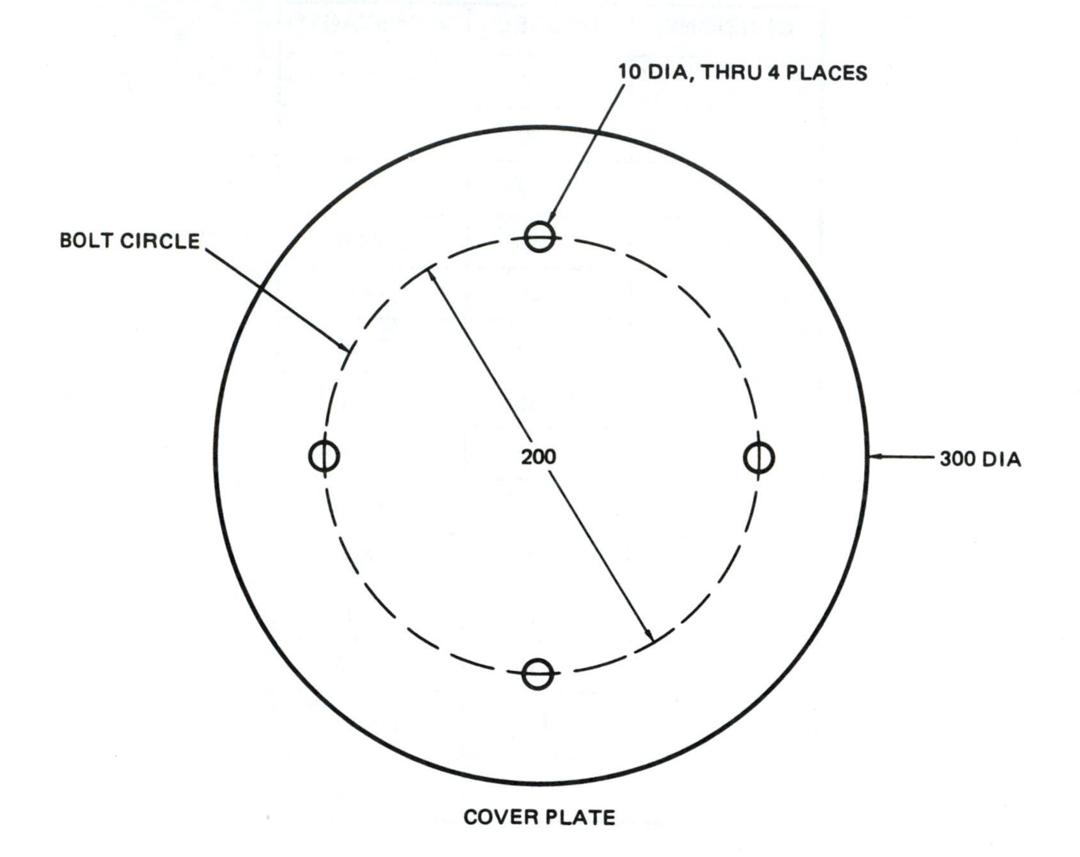

COVER PLATE

2. A bolt circle has a diameter of 475.5 mm. Around the circle are 32 holes, each 9 mm in diameter. How far apart are the centers of the holes? Round the answer to the nearer hundredth millimeter. ____________

3. Find the length of the chords used to make this adapter plate. Express each answer to the nearer hundredth millimeter.

   a. Chord *A* a.__________

   b. Chord *B* b.__________

   c. Chord *C* c.__________

   d. Chord *D* d.__________

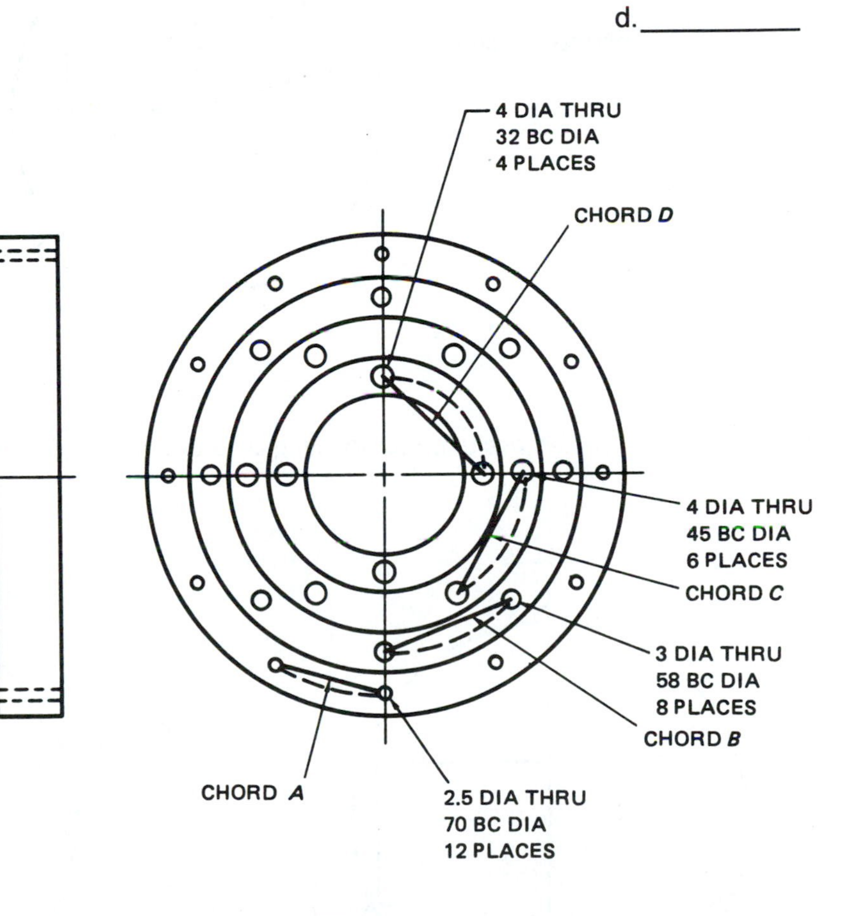

Note: Use this information for problems 4–6.

Measuring horizontal and vertical distances can be used to locate holes on a circle. The beginning point is the center of the circle. The distances are constants times the diameter of the circle.

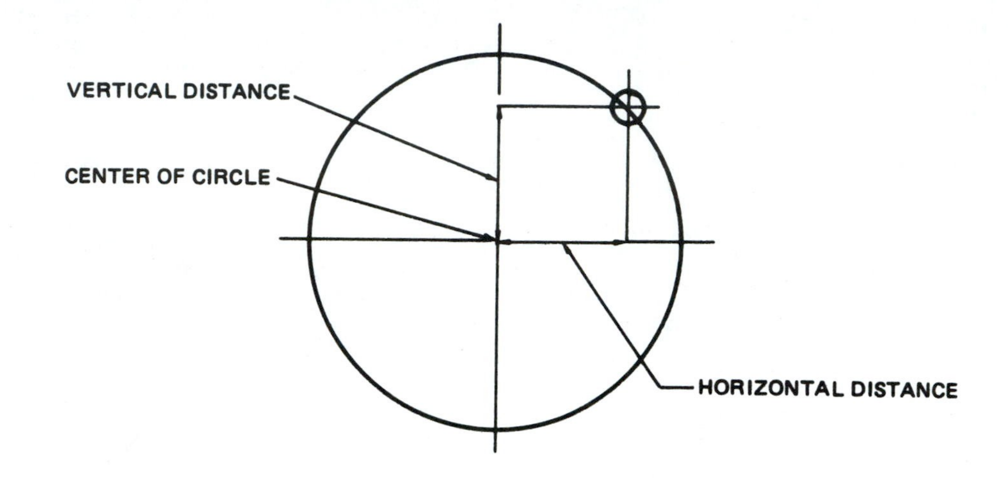

These constant values are used for locating different numbers of holes on a circle, where *D* equals the diameter.

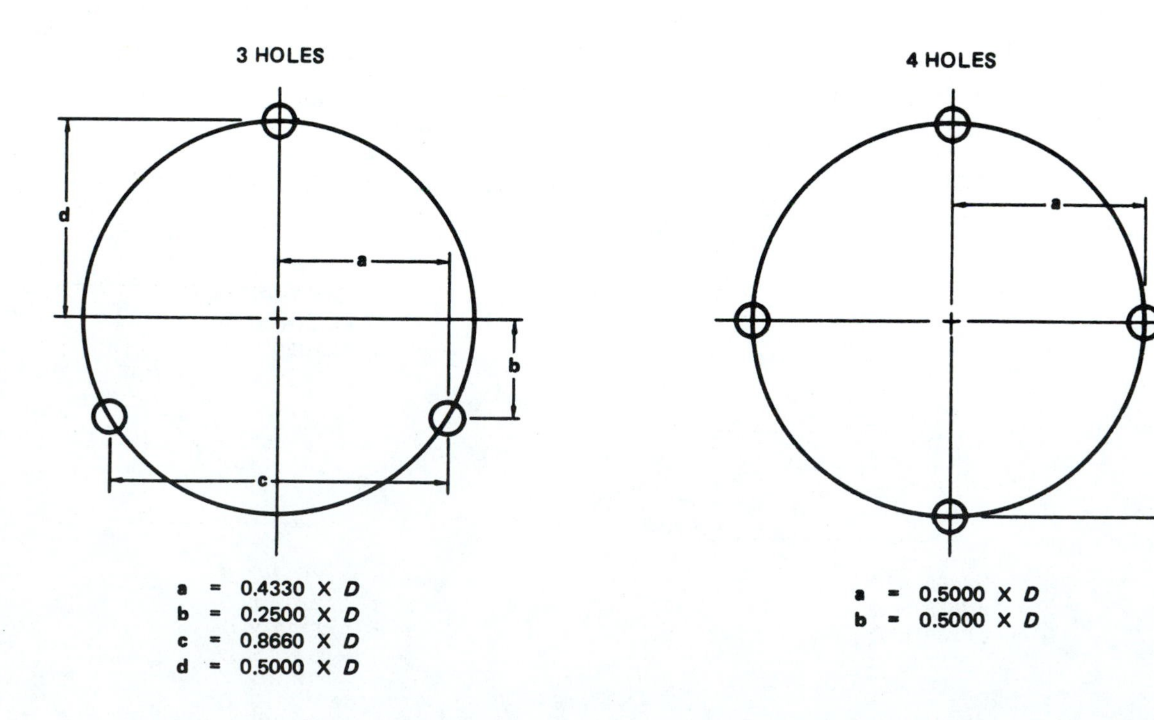

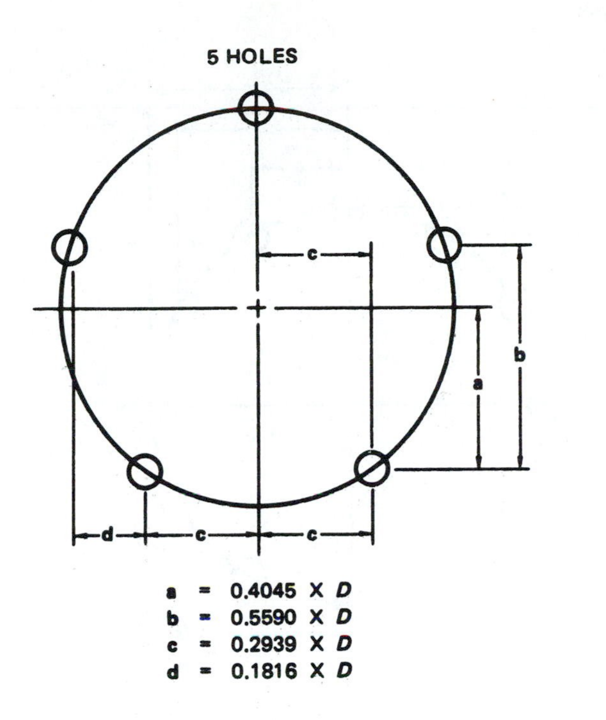
5 HOLES
a
b
c
c
d
c
a = 0.4045 X D
b = 0.5590 X D
c = 0.2939 X D
d = 0.1816 X D

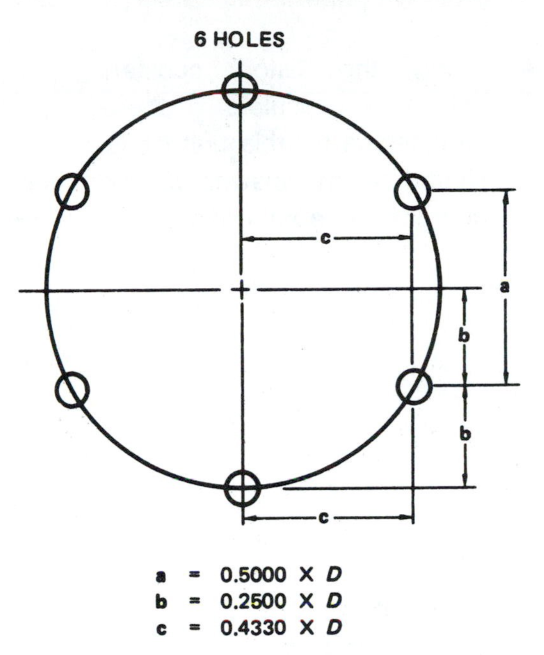
6 HOLES
c
a
b
b
c
a = 0.5000 X D
b = 0.2500 X D
c = 0.4330 X D

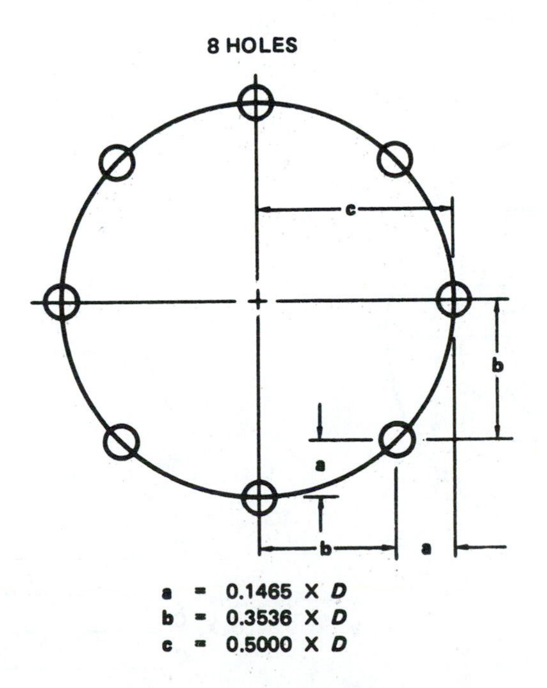
8 HOLES
c
b
a
b
a
a = 0.1465 X D
b = 0.3536 X D
c = 0.5000 X D

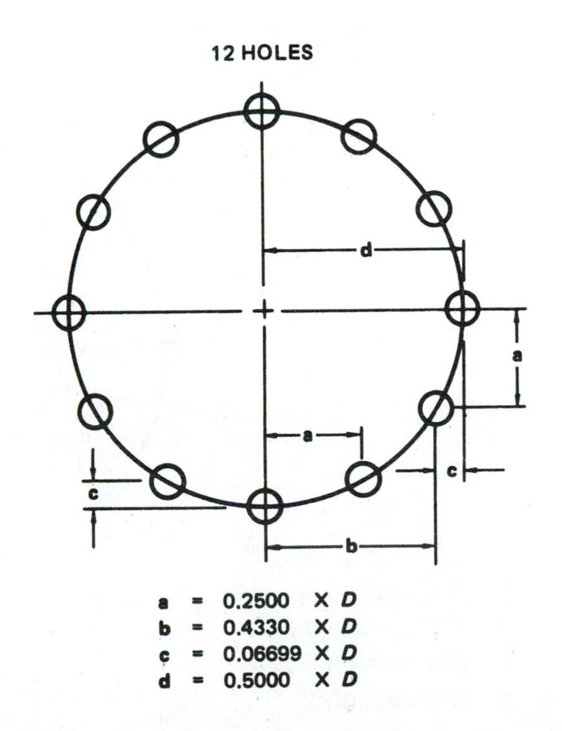
12 HOLES
d
a
a
c
c
b
a = 0.2500 X D
b = 0.4330 X D
c = 0.06699 X D
d = 0.5000 X D

4. Using the listed constant values, calculate these dimensions for this bolt circle. Round each answer to the nearer hundredth millimeter.

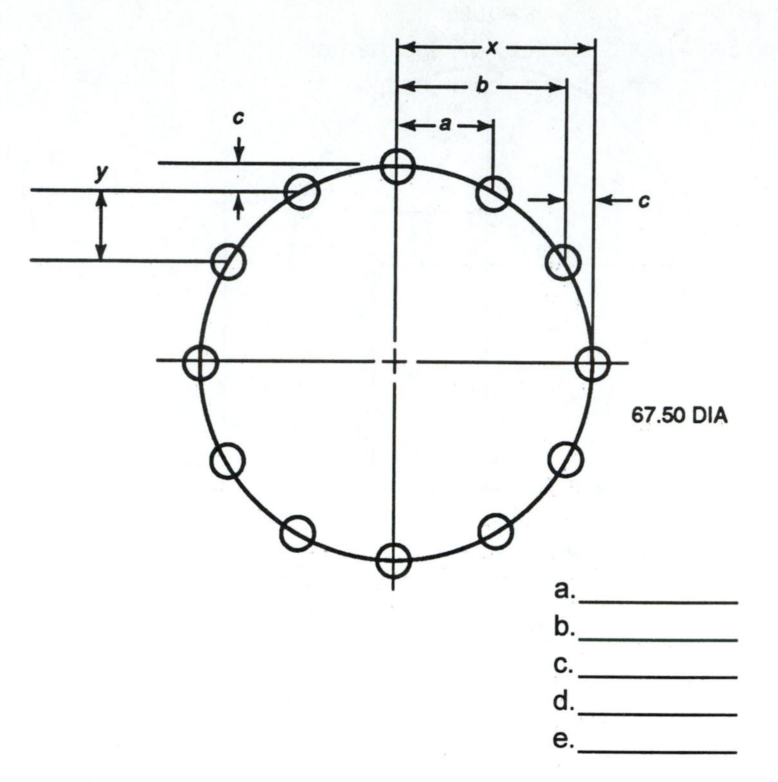

a. Dimension *a* a.__________
b. Dimension *b* b.__________
c. Dimension *c* c.__________
d. Dimension *x* d.__________
e. Dimension *y* e.__________

5. A cover plate must be made to fit a hole pattern. The diameter of the smaller circle is 4 inches. Find each dimension to the nearer thousandth inch.

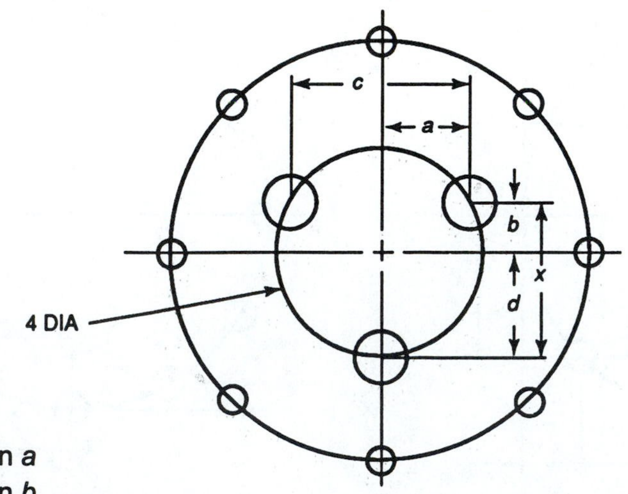

a. Dimension *a* a.__________
b. Dimension *b* b.__________
c. Dimension *c* c.__________
d. Dimension *d* d.__________
e. Dimension *x* e.__________

6. The diameter of the larger circle of the hole pattern is 8.250 inches. Find each dimension to the nearer thousandth inch.

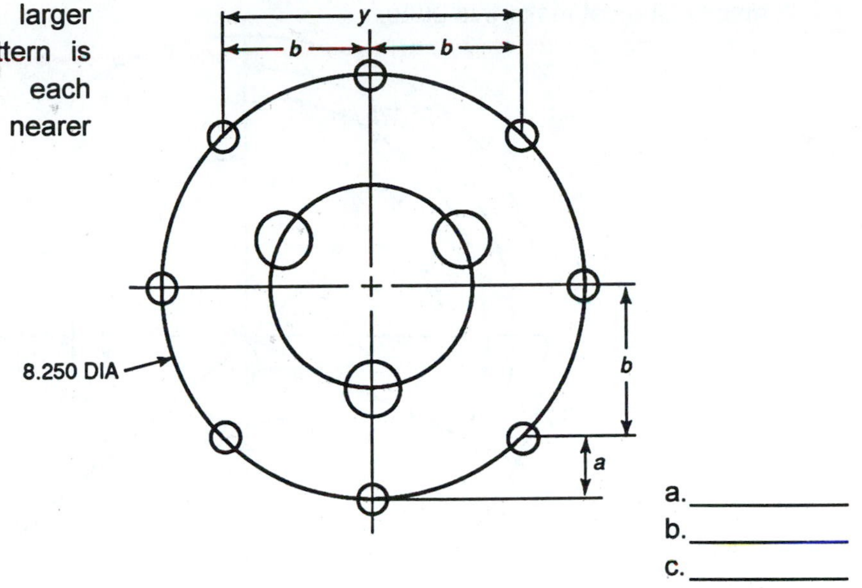

a. Dimension *a* a. ________
b. Dimension *b* b. ________
c. Dimension *y* c. ________

Note: Use this information for problems 7–9.

SECTION

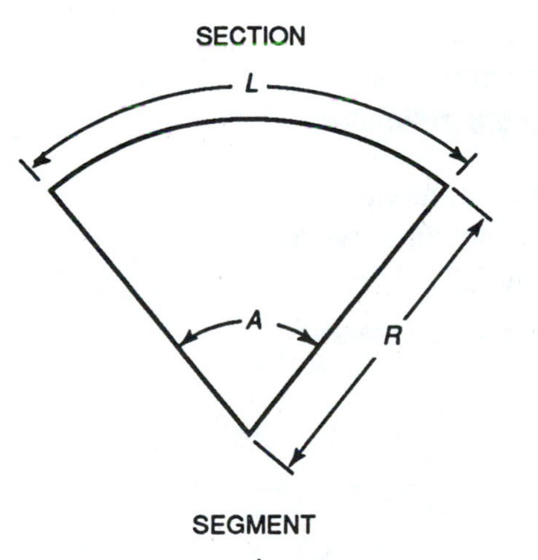

SEGMENT

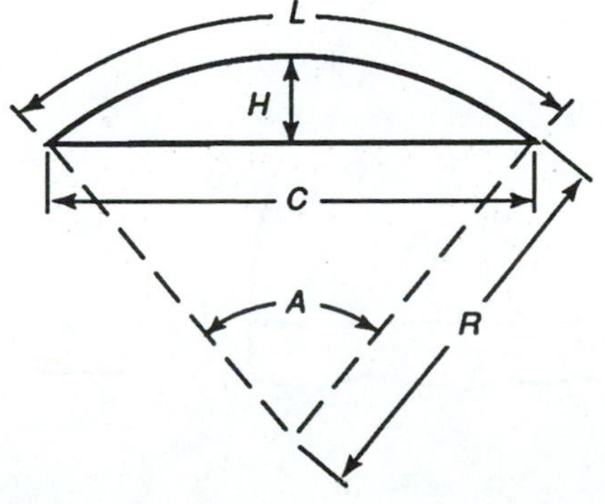

$$L = 0.01745 \times R \times A$$

$$R = \frac{57.296 \times L}{A}$$

$$A = \frac{57.296 \times L}{R}$$

$$R = \frac{L}{0.01745 \times A}$$

$$R = \frac{C^2 + 4H^2}{8H}$$

$$H = R - 0.5\sqrt{4R^2 - C^2}$$

$$C = 2\sqrt{H(2R - H)}$$

7. A machinist must make this plate.

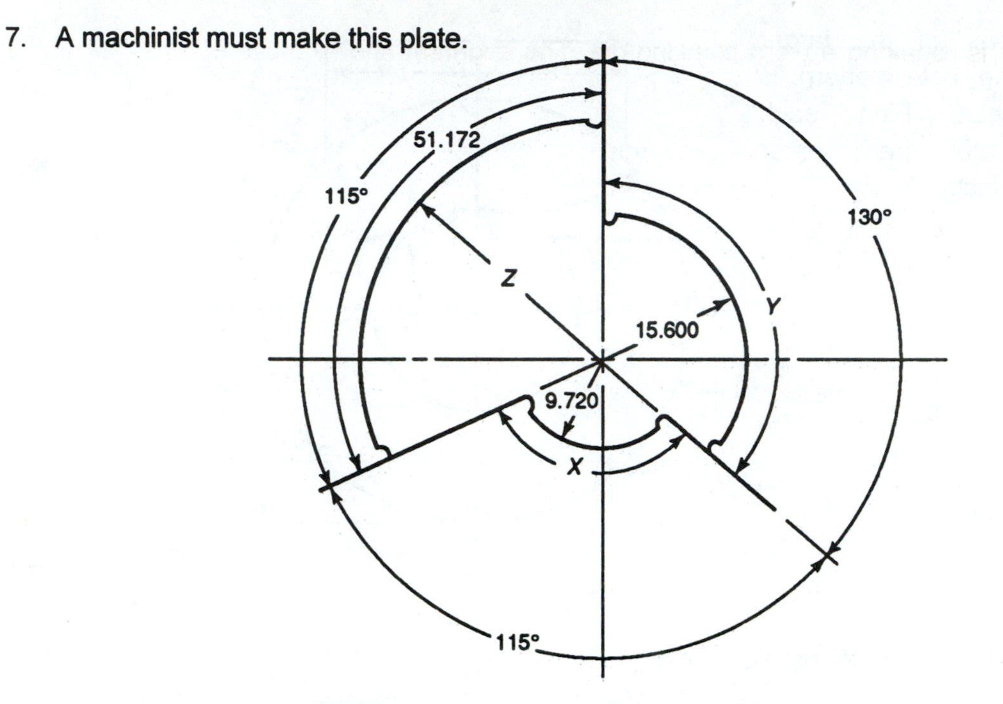

a. Find the length of arc *X* to the nearer thousandth millimeter. a.__________
b. Find the length of arc *Y* to the nearer thousandth millimeter. b.__________
c. Find the length of radius *Z* to the nearer thousandth millimeter. c.__________

8. When cutting keyways, the height of the chordal segment must be added to the depth of the keyway to insure a proper fit. Find the height of the chordal segment to the nearer hundredth millimeter. __________

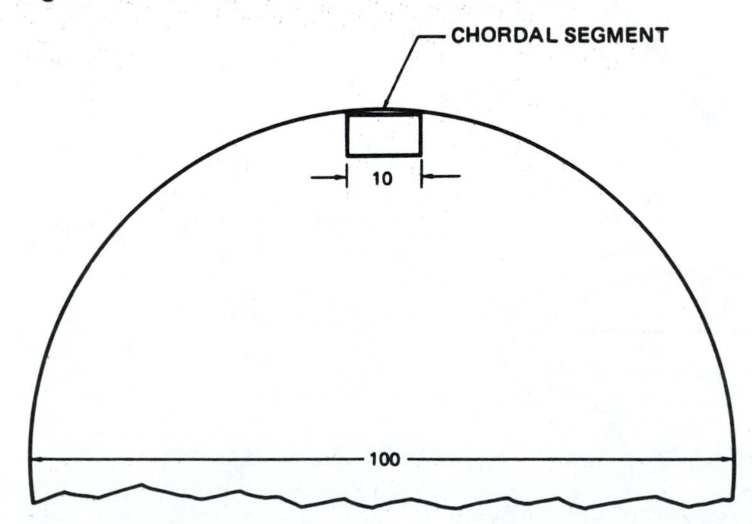

9. A diemaker is repairing a worn blanking die. The segment shown must be blanked.

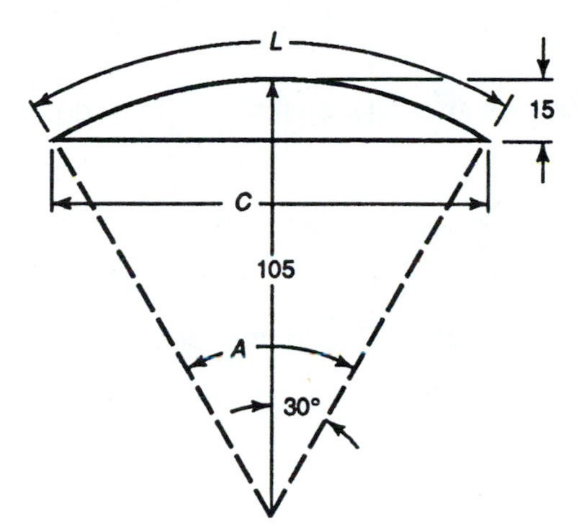

a. Find angle A. a.__________

b. Find the length of the segment (L) to the nearer hundredth millimeter. b.__________

c. Find the length of the chord (C) to the nearer hundredth millimeter. c.__________

Note: Use this information for problems 10–16.

$$1 \text{ radian (RAD)} = \frac{180^\circ}{\pi}$$

$$x \text{ radian} = x\left(\frac{180^\circ}{\pi}\right)$$

$$1^\circ = \left(\frac{\pi}{180^\circ}\right) \text{ radian}$$

$$x^\circ = x\left(\frac{\pi}{180^\circ}\right) \text{ radian}$$

10. How many degrees are π radians? __________

11. One degree is equal to how many radians? Round the answer to five decimal places. __________

12. One radian is equal to how many degrees? Round the answer to four decimal places. __________

13. Express 45° as radians. __________

14. Express 2 radians as degrees. Round the answer to four decimal places. __________

15. How many radians are in 143 degrees? Round the answer to five decimal places. __________

16. How many degrees are in 1.5 radians? Round the answer to four decimal places. __________

Unit 41 PERIMETERS AND BEND ALLOWANCES

BASIC PRINCIPLES

- *Perimeters* are important to the die maker and press room personnel because a perimeter, along with the thickness and type of material, determines how large a punch press must be used to blank a part. If the press used is too small, the press may be damaged.

- *Bend allowances* are important in the press room when bending metal parts. When a material is bent in a press brake, stretching, compression, and the radius of a bend must be taken into account so that the final product is the proper size. The following problems provide you with the practice needed to understand these important parameters.

PRACTICAL PROBLEMS

1. Find the perimeter of this cover plate to the nearer hundredth millimeter. __________

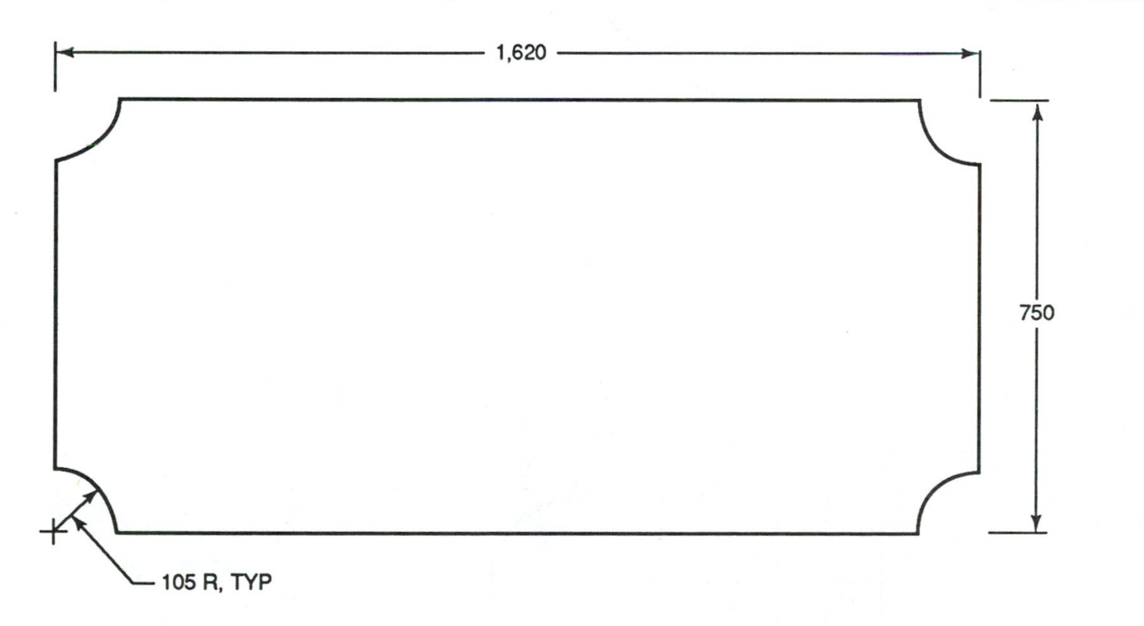

2. A die maker must make this part. What is the perimeter to the nearer hundredth millimeter? ______

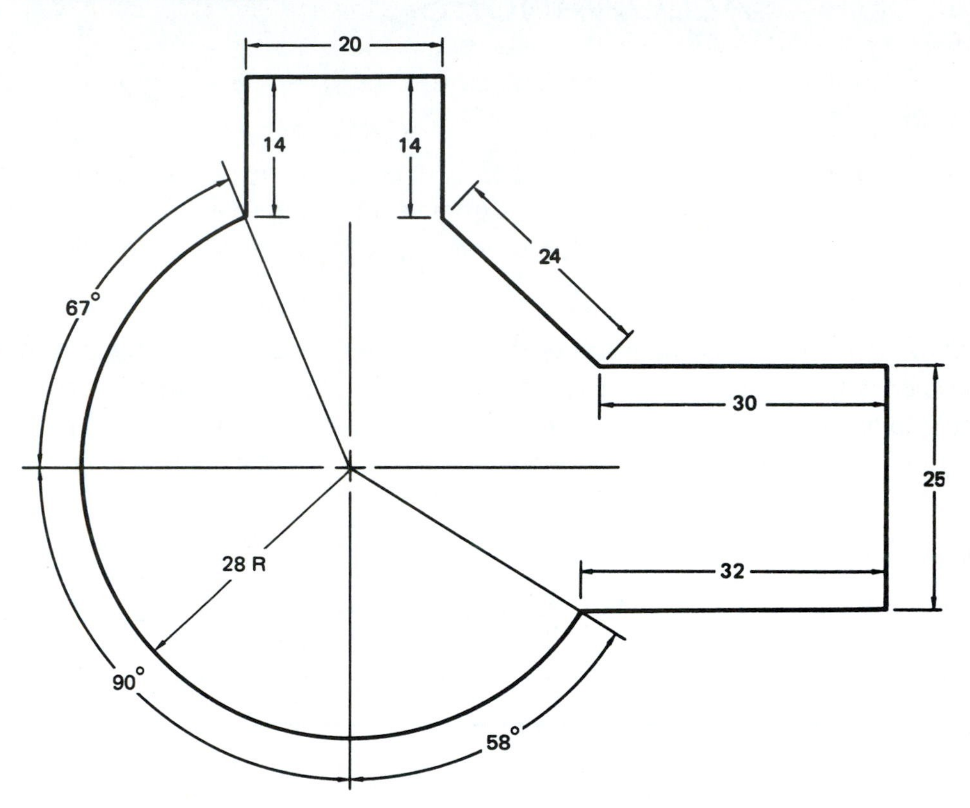

3. Find, to the nearer hundredth millimeter, the perimeter of this die opening. ______

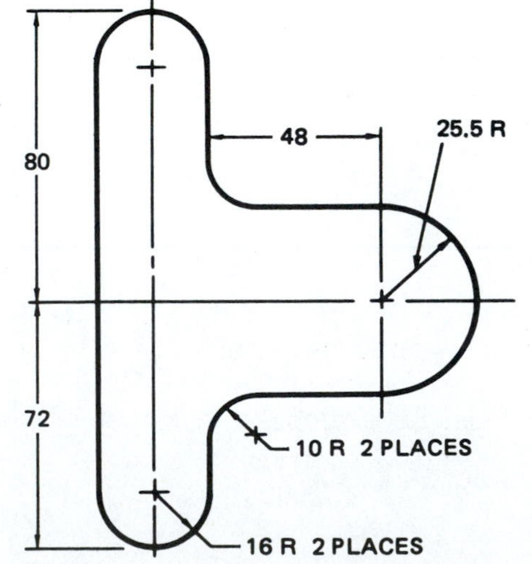

Note: Use this information for problems 4–6.

When calculating blanking pressure for a stamping operation, the die maker must know the perimeter of the part, the thickness of the metal, and the metal shear strength. This chart and formula are used to calculate the blanking pressure for metal stampings.

APPROXIMATE SHEAR STRENGTH		
MATERIAL	SHEAR STRENGTH	
	lb./sq. in.	kg/mm^2
Low Carbon Steel *1020*	43,000	30.23
Med Carbon Steel *1050*	81,000	56.94
High Carbon Steel *1095*	105,000	73.82
Alloy Steels		
2340	123,000	86.47
3250	140,000	98.42
4130	80,000	56.24
Aluminum Alloys	28,000	19.68
Brass	35,000	24.61
Magnesium Alloys	25,000	17.58

SYMBOL	MEANING
BP	blanking pressure
P	perimeter of the part
t	thickness of the part
s	shear strength

$$BP = P \times t \times s$$

The blanking pressure is measured in tons or megagrams. One megagram is equal to 1,000 kilograms. One megagram is sometimes referred to as one metric tonne.

Note: The perimeter and thickness must be expressed in inches or millimeters. The blanking pressure is $\frac{\text{kilogram}}{1,000}$ or $\frac{\text{pounds}}{2,000}$

4. This part is made from 0.5-mm = thick *4130* steel. What is the pressure required to stamp the part? Round the answer to the nearer thousandth megagram. ____________

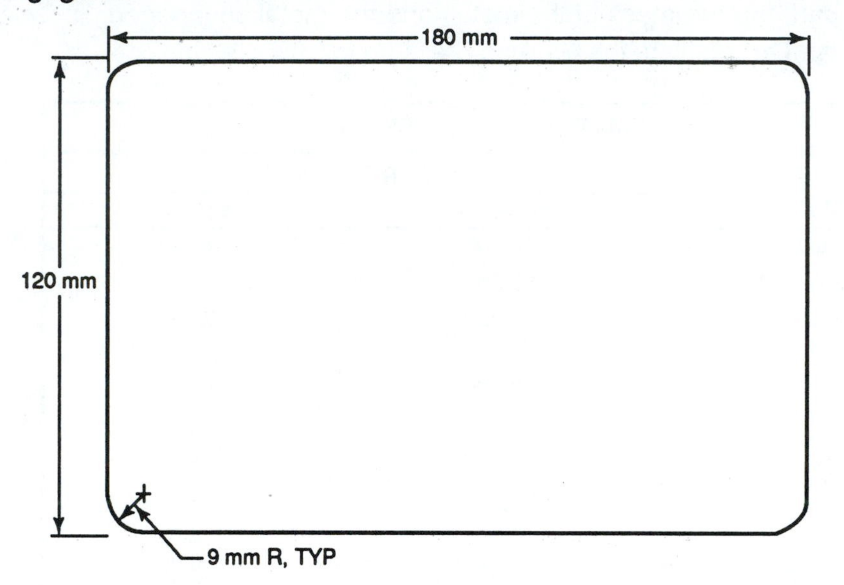

5. A die maker has three punch presses that can be used to stamp parts. Press *#1* has a capacity of 25 megagrams; press *#2* has a capacity of 40 megagrams; press *#3* has a capacity of 50 megagrams. Complete this table to determine the press that should be used on each job.

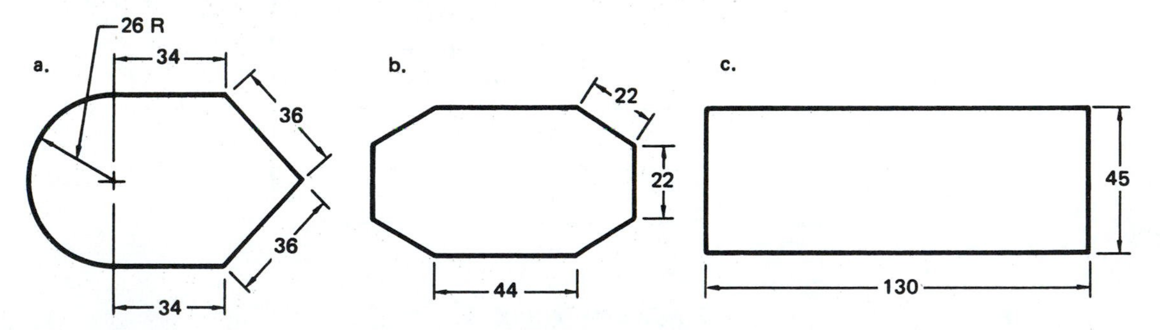

MATERIAL	a. *1095* steel	b. *1020* steel	c. magnesium alloy
THICKNESS	1.5 mm	7 mm	5 mm
PERIMETER			
SHEAR STRENGTH			
BLANKING PRESSURE (to the nearer megagram)			
PRESS NUMBER			

6. This panel cover is made from 0.156-inch-thick *1020* steel. What is the blanking pressure to the nearer tenth ton? ____________

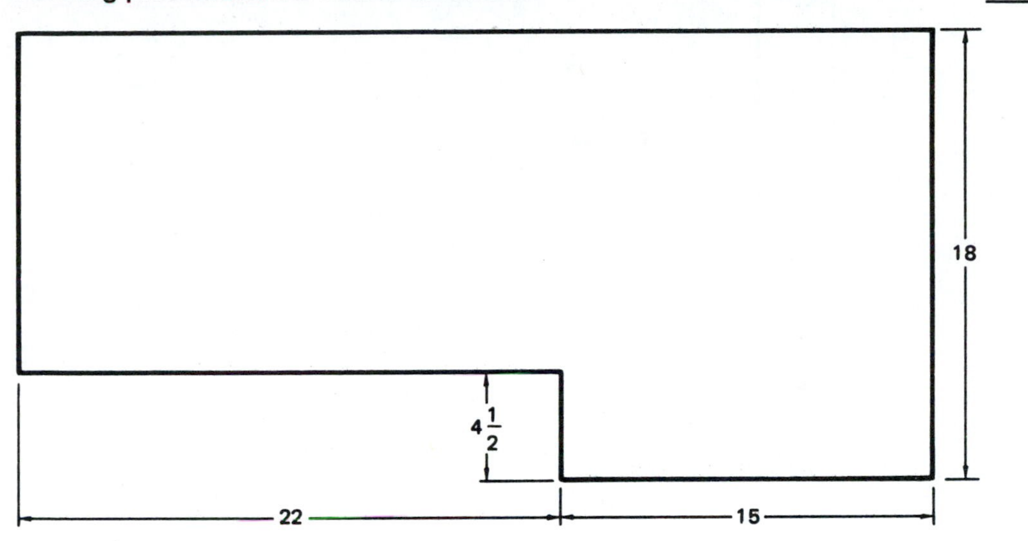

Note: Use this information for problems 7–8.

When metal is bent, a portion of the metal expands while the rest compresses.

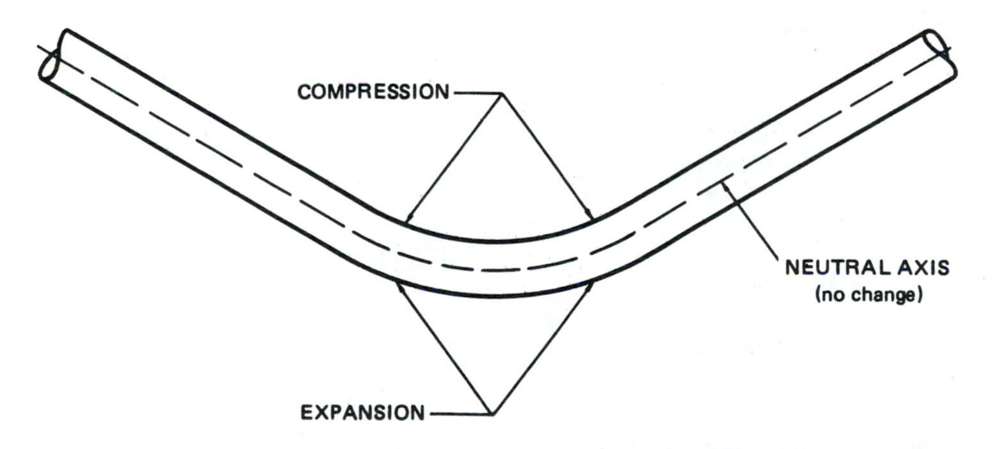

To insure the desired length is obtained after bending, a bend allowance is found. The total of the bend allowance plus the length of the straight piece gives the needed length of the blank.

This formula is used for finding the bend allowance.

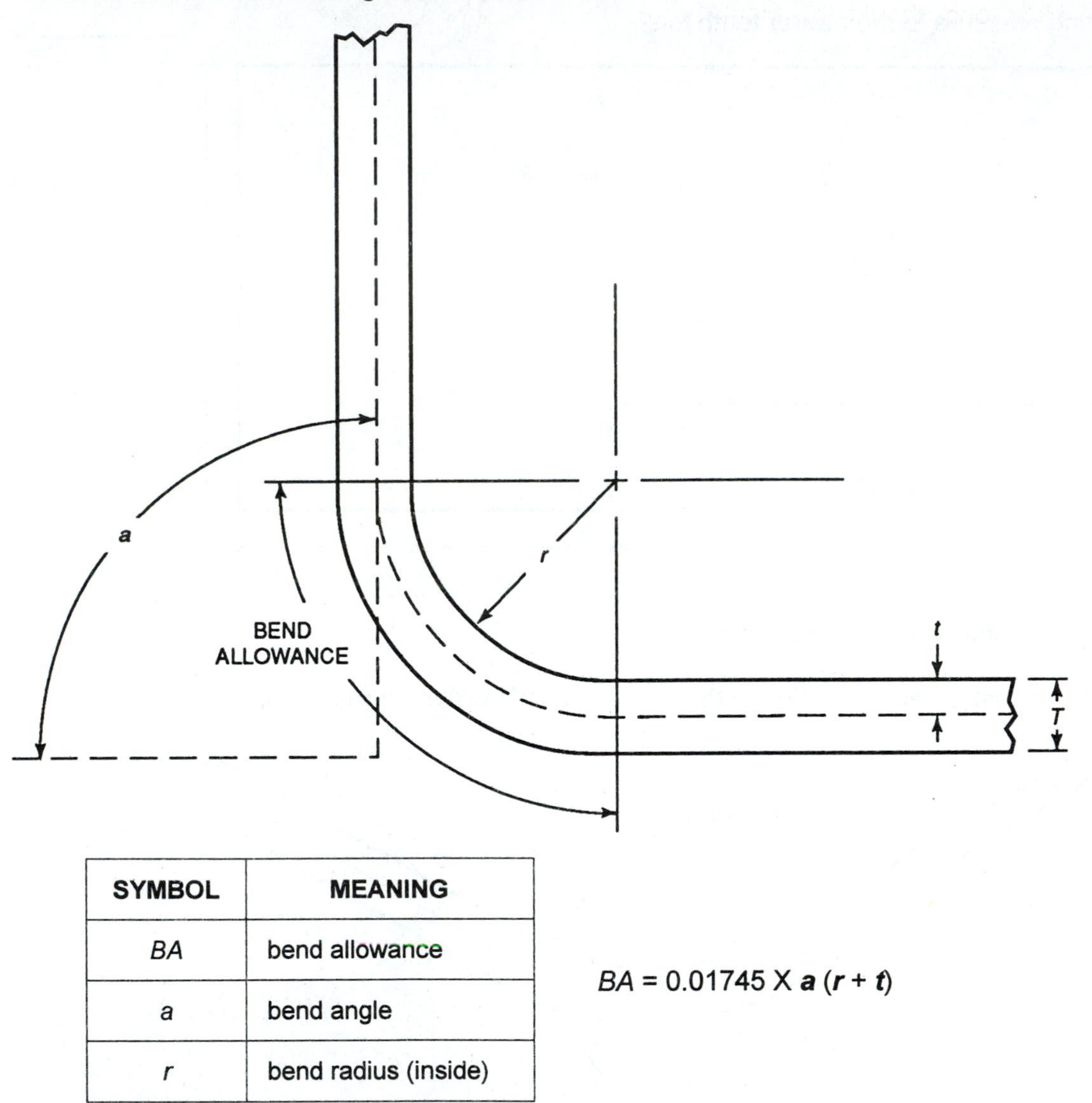

SYMBOL	MEANING
BA	bend allowance
a	bend angle
r	bend radius (inside)

$BA = 0.01745 \times \mathbf{a} (\mathbf{r} + \mathbf{t})$

The thickness ***t*** has different values depending upon the bend radius.

If the bend radius ***r*** is equal to or more than twice the metal thickness, ***t*** = $\frac{1}{2}$ **T** *or* **t** = 0.5**T**.

If the bend radius ***r*** is less than twice the metal thickness, ***t*** = $\frac{1}{3}$ **T** *or* **t** = .033**T**.

The bend angle is the number of degrees that the material is actually bent.

7. This lever is made from 15-mm-thick stock. How long should the blank be before bending? Express the answer to the nearer hundredth millimeter. ______

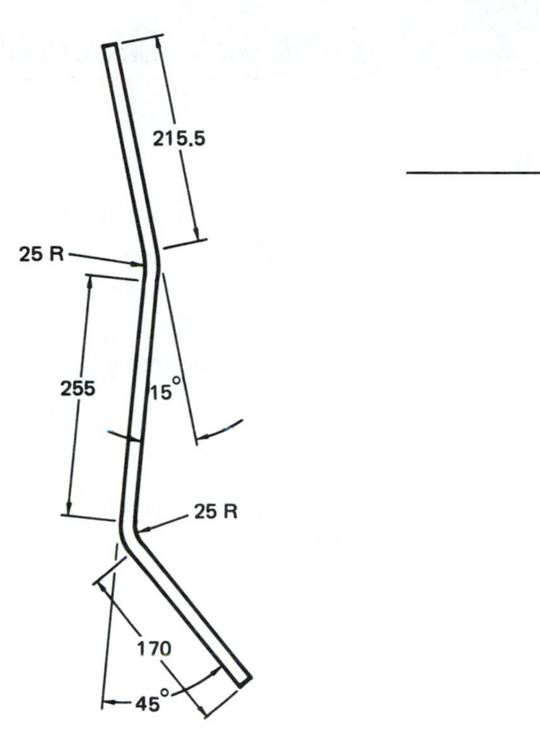

8. Two parts are to be made.

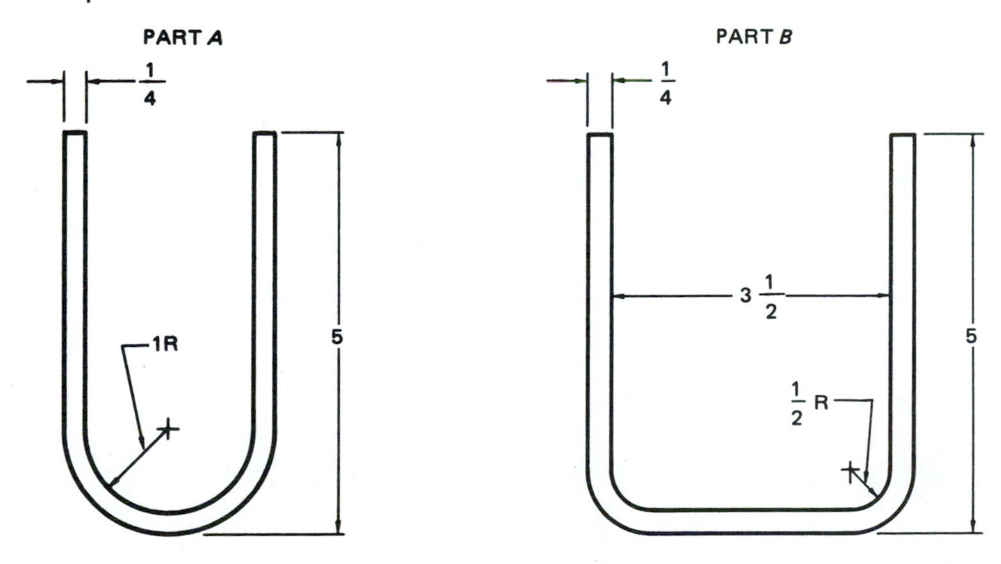

a. Find, to the nearer thousandth inch, the length of blank needed for part *A*. a. ______

b. Find, to the nearer thousandth inch, the length of blank needed for part *B*. b. ______

Unit 42 GEOMETRIC CONSTRUCTION

PRINCIPLES

- An understanding of geometric construction is important in the manufacturing field. Die making, for example, requires a machinist to lay out geometric figures on a daily basis. The greater the understanding of geometric construction the die maker has, the faster and more accurately the project can be completed. Discuss the various techniques of geometric construction with your instructor and complete the following problems.

PRACTICAL PROBLEMS

(Use a separate sheet of paper.)

1. Construct a line segment parallel to $\overline{AB}$ and 30 mm above it.

   A ———————————————— B

2. Bisect line segment $\overline{CD}$.

   C ———————————————— D

3. Construct a line segment perpendicular to $\overline{EF}$ at point *x*.

   E ——————●—————————— F
   x

4. Construct a line segment perpendicular to $\overline{GH}$ and passing through point *y*.

   ● y

   G ———————————————— H

5. Construct an angle equal to angle ***ABC***.

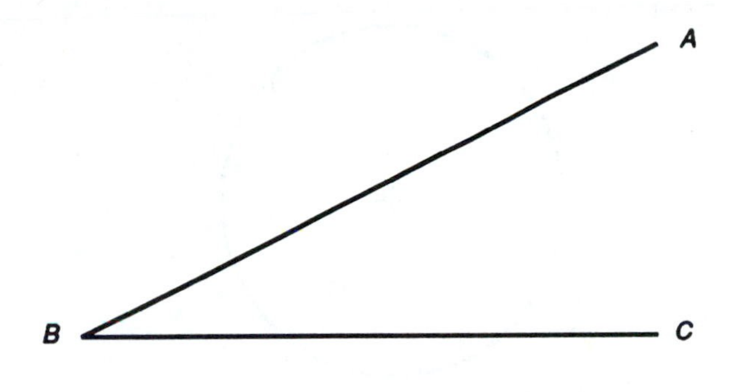

6. Bisect angle ***DEF***.

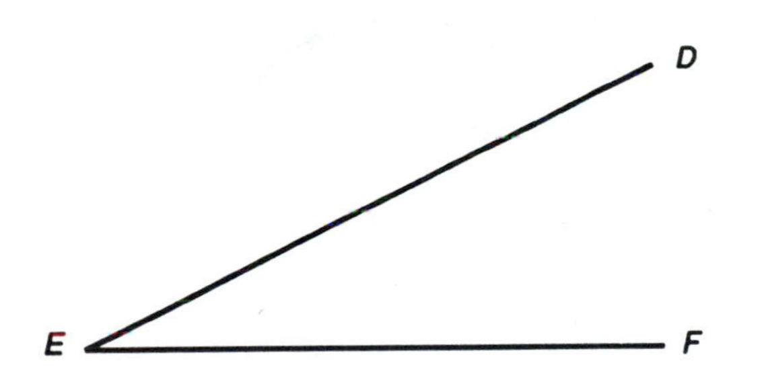

7. Divide this line segment into 7 equal parts.

8. Find the center of this circle.

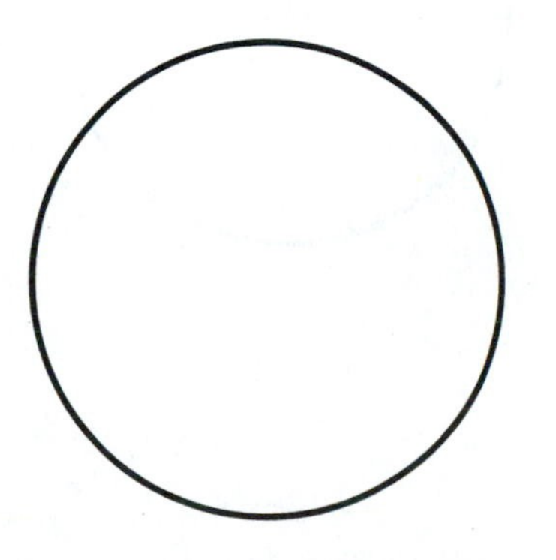

9. Construct a square inside this circle.

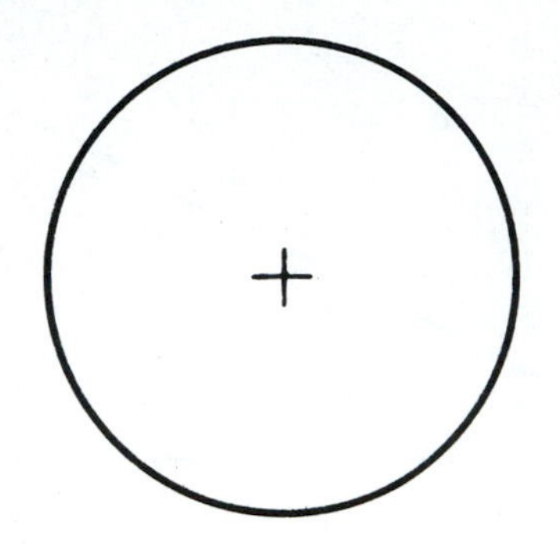

10. Construct a hexagon inside this circle.

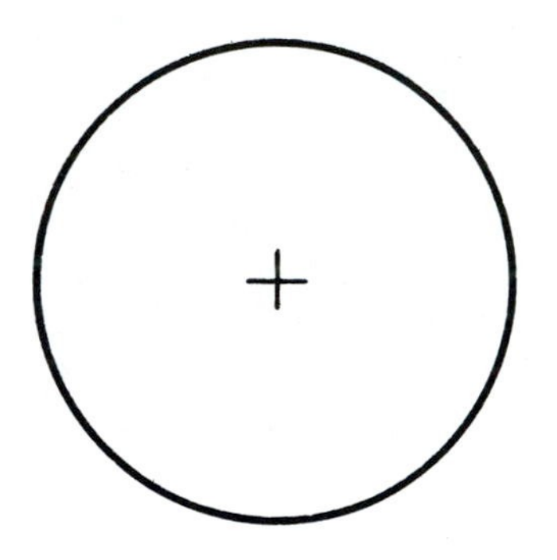

11. Construct an octagon inside this circle.

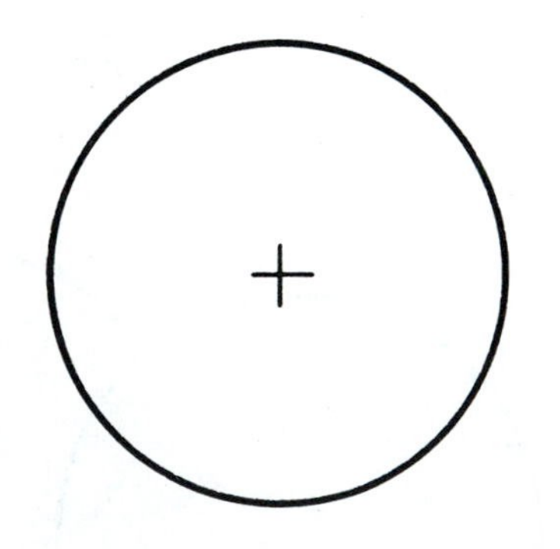

12. Circumscribe a circle outside each shape.

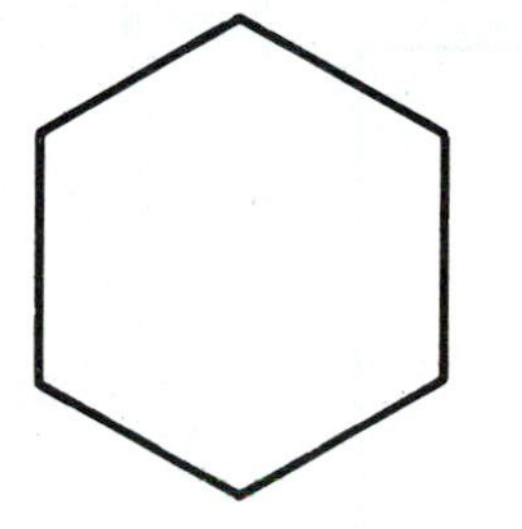

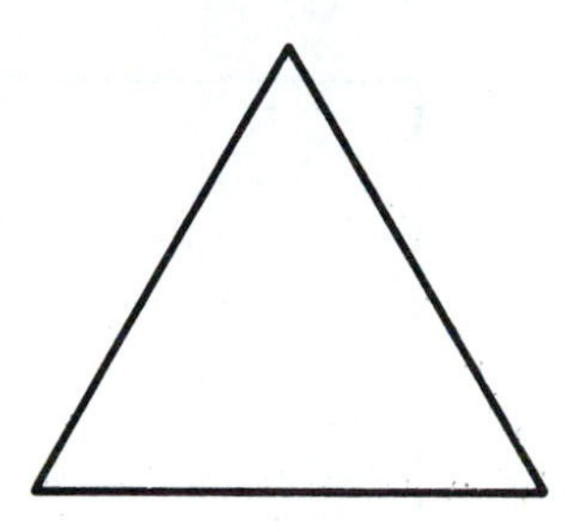

14. Inscribe a circle inside each shape.

14. Construct a ray parallel to $\overline{AB}$ and 40 mm above it. The end point of the ray should be 40 mm from the center of the circle.

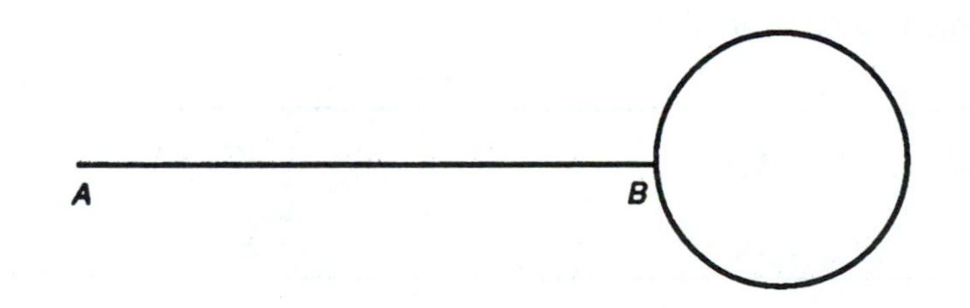

15. Using a 12-mm radius, construct arcs at the four corners of this part.

16. Locate the center of this segment and find its diameter in inches.

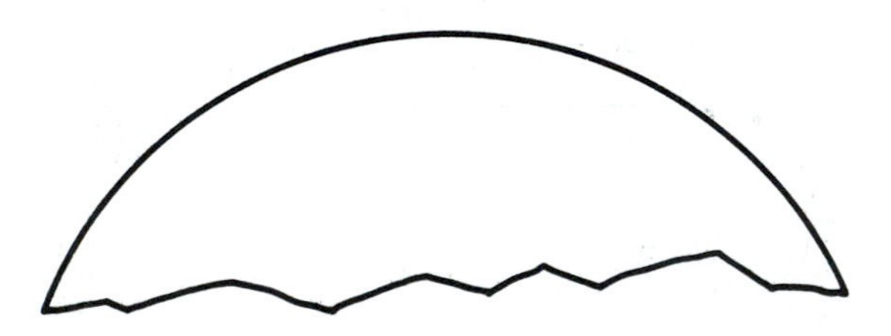

17. Ten equally spaced holes are punched along the edge of a piece of stock. The distance between the centers of the extreme holes is 127 mm. Lay out the holes by construction.

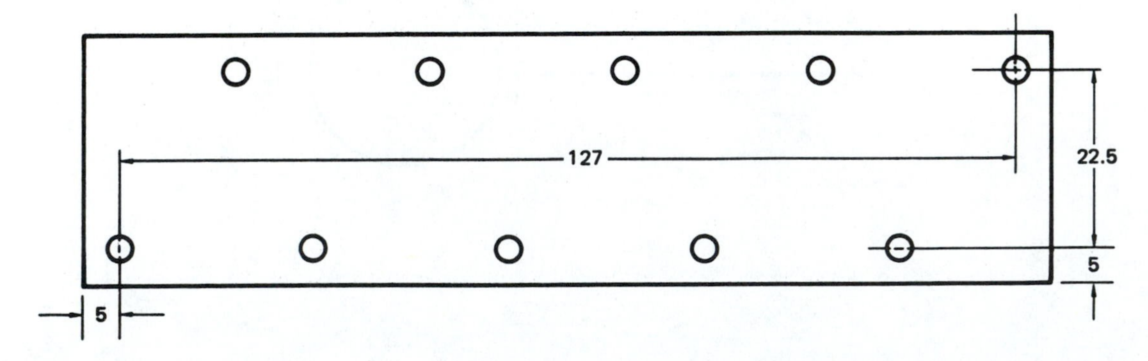

CRITICAL THINKING PROBLEMS

1. Determine the *x, y* coordinates of holes 1 through 8 relative to point *A*. The bolt circle diameter is 3.4901″.

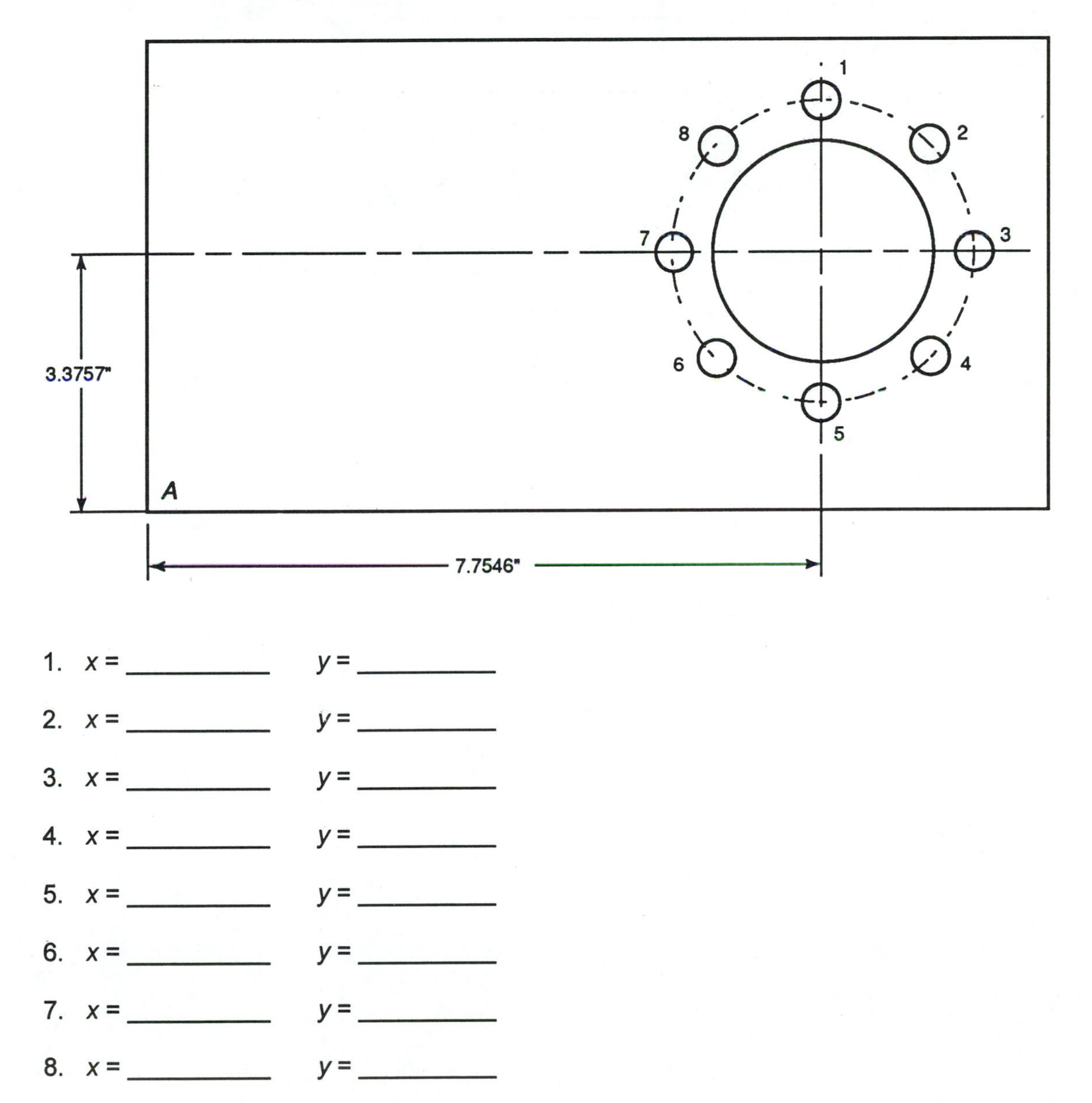

1. *x* = __________ *y* = __________

2. *x* = __________ *y* = __________

3. *x* = __________ *y* = __________

4. *x* = __________ *y* = __________

5. *x* = __________ *y* = __________

6. *x* = __________ *y* = __________

7. *x* = __________ *y* = __________

8. *x* = __________ *y* = __________

2. Determine the force required to blank the following part. The material used is *1020* steel. The thickness is #7 gage.

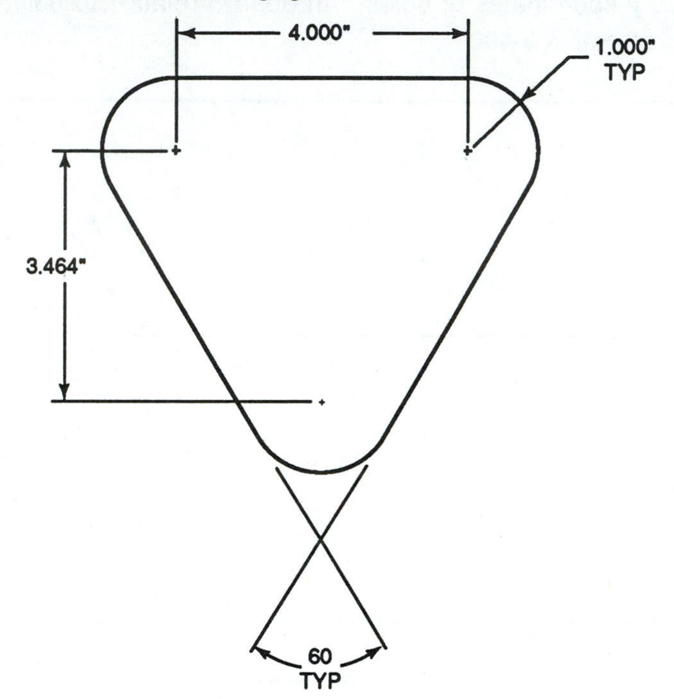

3. Determine the weight of the part in question #2.

Trigonometry

SECTION 11

Unit 43 NATURAL FUNCTIONS

BASIC PRINCIPLES

- There are six functions used in trigonometry. These trigonometric functions are actually the ratios between the three sides of a triangle. The ability to determine the values of the six functions is necessary in the study of trigonometry and in daily machine shop operations. The value of a trigonometric function can be extracted from tables published in various handbooks as in the appendix of this book. A much faster way of determining a function is to use the scientific calculator. This is the method most used today; however, the ability to find values from the *Tables of Trigonometric Functions* may be necessary at some time. The following problems will provide practice in determining the value of the functions of various angles.

Example: Use the Tables of Trigonometric Functions in the Appendix to determine the sine of a 10° angle.

Step 1: Locate 10° along the left edge of the table.

Step 2: Read right and locate the Sin column at the top of the table.

Step 3: Read .17365 in the 10° row and the Sin column.

Example: Use the Tables of Trigonometric Functions in the appendix to determine the tangent of a 73° angle.

Step 1: Locate 73° along the right edge of the table (Note that above 45° the angle is found on the right edge of the table and the column headings are found at the bottom of the table.)

Step 2: Locate the Tan column at the bottom of the table.

Step 3: Read the Tan of the angle in the 73° row and the Tan (bottom) column. The tan of 73° = 3.2709.

Generally, an angle is not an even angle as in the last two examples. It is possible to read directly functions of angels at 10′ increments with the table provided in the appendix.

Example: Use the Tables of Trigonometric Functions in the appendix to determine the cosine of a 54°20′ angle.

Step 1: Locate 54° along the right edge of the table.

Step 2: Continue reading upward until 20′ is found.

Step 3: Locate the Cos column (bottom) and read the cos of 54°20′. The cos of 54°20′ = .58307.

It is often necessary to determine the function of an angle measured in minutes and seconds that is not in the table. An example is 33°25′ whose functions lie between the functions of 33°20′ and 33°30′ .

Example: Use the Tables of Trigonometric Functions in the appendix to calculate the sine of 33°25′.

Step 1: Locate the sin of 33°20′ and 33°30′.

$$\sin 33°20' = .54951$$

$$\sin 33°30' = .55194$$

Step 2: Calculate the difference between the two functions.

$$\begin{array}{r} .55194 \\ -\ \underline{.54951} \\ .00243 \end{array}$$

Step 3: Since 25′ is halfway between 20′ and 30′, take 5⁄10 of the difference and add it to the function of 33°20′.

$$\frac{5}{10} \text{ of } .00243 = .001215$$

$$\text{Sin of } 33°20' = .54951$$

$$\begin{array}{r} .54951 \\ +\ \underline{.00122} \\ .55073 \end{array}$$

The sin of 33°25′=.55073

Many angles are used that have accuracy in seconds. It is necessary to determine the function of these angles through calculation.

Example: Use the Tables of Trigonometric Functions in the appendix to calculate the tangent of 11°48′14″.

Step 1: Since 11°48′14″ falls between 11°40′ and 11°50′ in the table, locate the functions of these angles.

$$\tan 11°40' = .20648$$
$$\tan 11°50' = .20952$$

Step 2: Calculate the difference between the two functions.

$$\begin{array}{r} .20952 \\ -\ \underline{.20648} \\ .00304 \end{array}$$

Step 3: Take $\frac{8}{10}$ of the difference and add it to the function of 11°40′.

$$\frac{8}{10} \text{ of } .00304 = .00243$$
$$\text{Tan of } 11°40' = .20648$$
$$+\ \underline{.00243}$$
$$\text{Tan of } 11°48' = .20891$$

Step 4: Calculate the function of the seconds portion of the angle.

$$\frac{14}{600} \text{ of } .00304 = .00007$$

Step 5: Add this to the function of 11°48′:

$$\text{Tan } 11°48' = .20891$$
$$+\ \underline{.00007}$$
$$\text{Tan } 11°48'14'' = .20898$$

- From the previous examples it can be seen that using trigonometric tables to find functions of angles is very laborious and error prone. The scientific calculator eliminates these problems by requiring only that the angle be entered in decimal degrees. The key with the function needed is then pressed, and the function simply appears on the display. To convert an angle using minutes and seconds to decimal degrees, the following method may be used.

Example: Convert 11°48′14″ to decimal degrees.

Step 1: Divide seconds by 60.

$$14 \div 60 = 0.23333$$

Step 2: Add to minutes.

$$48 + 0.23333 = 48.23333$$

Step 3: Divide by 60.

$$48.23333 \div 60 = .80388$$

Step 4: Add to degrees.

$$11^\circ + 0.80388 = 11.80388^\circ$$

PRACTICAL PROBLEMS

1. Complete this chart for the values of the trigonometric functions. Round answer to the nearest one thousandth.

	FUNCTION	ANGLE	VALUE
a.	sine	64°	
b.	sine	86° 6′	
c.	cosine	41° 22′	
d.	tangent	26° 14′	
e.	cosine	52° 58′	
f.	tangent	72° 7′	

2. Complete this chart for the angles of each function.

	FUNCTION	ANGLE	VALUE
a.	sine		0.42262
b.	tangent		0.38386
c.	cosine		0.73135
d.	tangent		2.05030
e.	sine		0.77715
f.	cosine		0.10453

3. Complete this chart for the values of the functions.

	FUNCTION	ANGLE	VALUE
a.	sine	21° 13′ 17″	
b.	sine	21° 13′ 52″	
c.	tangent	14° 18′ 27″	
d.	cotangent	56° 49′ 52″	
e.	cosine	10° 13′ 16″	
f.	sine	82° 43′ 42″	
g.	cotangent	61° 7′ 14″	
h.	cosine	29° 13′ 42″	
i.	tangent	28° 29′ 26″	
j.	cotangent	66° 22.6′	

4. Express each angle in degrees, minutes, and seconds and find the values of the trigonometric functions.

	FUNCTION	ANGLE (in degrees)	ANGLE (in degrees, minutes, and seconds)	VALUE
a.	sine	26.25°		
b.	cosine	81.36°		
c.	tangent	61.08°		
d.	cosine	7.92°		
e.	sine	89.66°		
f.	tangent	22.46°		

Unit 44 RIGHT TRIANGLES

BASIC PRINCIPLES

- A *right triangle* is one having a right angle. A *right angle* is an angle of 90°. The figure shows a right triangle and the standard labeling used to identify the various angles and sides of a right triangle.

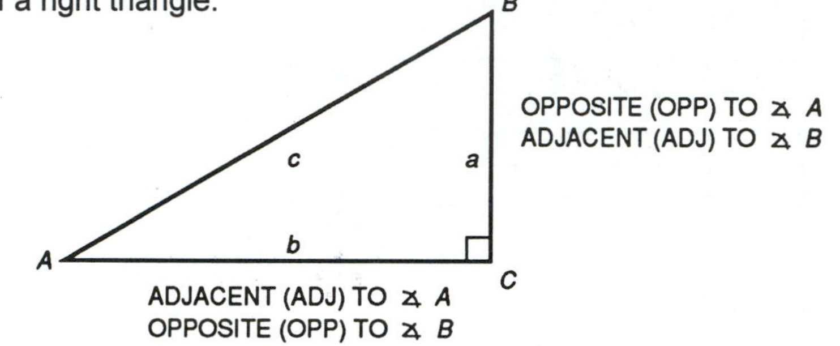

The six trigonometric functions used in right triangle trigonometry are as follows:

$$\text{SINE } A = \frac{\text{OPP}}{\text{HYP}} = \frac{a}{c} \qquad \text{COSECANT } A = \frac{\text{HYP}}{\text{OPP}} = \frac{c}{a}$$

$$\text{COSINE } A = \frac{\text{ADJ}}{\text{HYP}} = \frac{b}{c} \qquad \text{SECANT } A = \frac{\text{HYP}}{\text{ADJ}} = \frac{c}{b}$$

$$\text{TANGENT } A = \frac{\text{OPP}}{\text{ADJ}} = \frac{a}{b} \qquad \text{COTANGENT } A = \frac{\text{ADJ}}{\text{OPP}} = \frac{b}{a}$$

With these functions, if any two sides or one side and one of the acute angles is known, the other sides and angles can be found. It is helpful to memorize these functions as they are an important part of manufacturing. Discuss right angle trigonometry with your instructor and complete the following problems.

PRACTICAL PROBLEMS

Note: Use this diagram for problems 1–4.

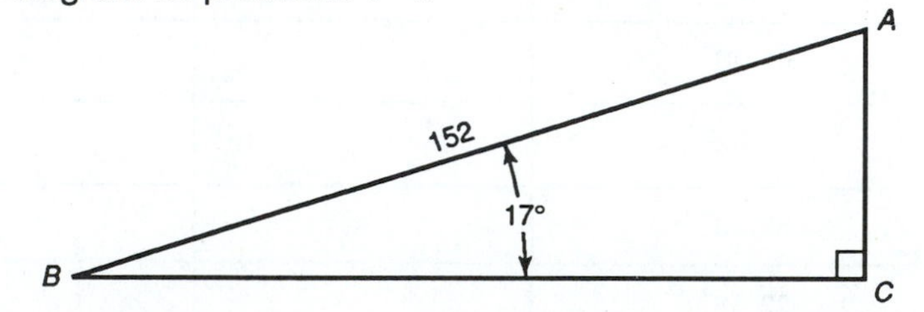

1. Locate side ***AC*** on the right triangle.

   a. What function is used to find side ***AC*** relative to angle ***B***? a.__________

   b. Find the length of side ***AC*** to the nearer hundredth millimeter. b.__________

2. Locate side ***BC*** on the right triangle.

   a. What function is used to find side ***BC***? a.__________

   b. Find the length of side ***BC*** to the nearer hundredth millimeter. b.__________

3. If angle ***B*** is changed to 42 degrees, find the length of side ***AC*** to the nearer hundredth millimeter. __________

4. If angle ***B*** is changed to 42 degrees, find the length of side ***BC*** to the nearer hundredth millimeter. __________

Note: Use this diagram for problems 5 and 6.

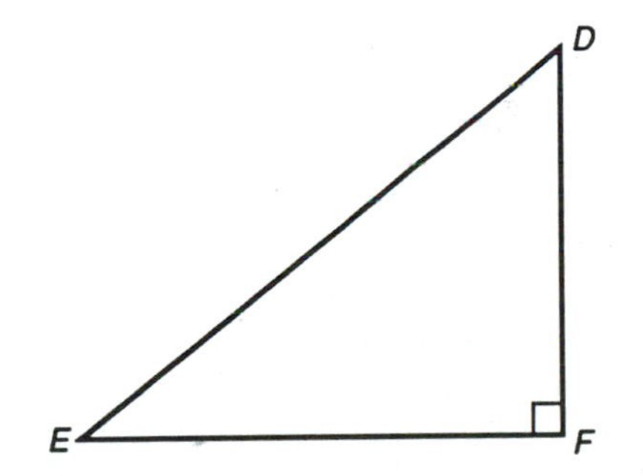

5. In triangle ***DEF***, side ***DE*** is 3 inches and angle ***E*** is 37 degrees. Find side ***DF*** to the nearer hundredth inch. __________

6. In triangle ***DEF***, side ***DE*** is 152.4 millimeters and angle ***E*** is 22 degrees 40 minutes. Find side ***EF*** to the nearer hundredth millimeter. __________

7. Two holes are to be located in a plate as shown. After boring a hole at point ***B***, the table carrying the plate is moved horizontally the distance ***Y*** and vertically the distance ***X***. This locates the hole to be bored at point ***A***. Find the values of ***X*** and ***Y*** for each set of measurements. Round each answer to the nearer thousandth.

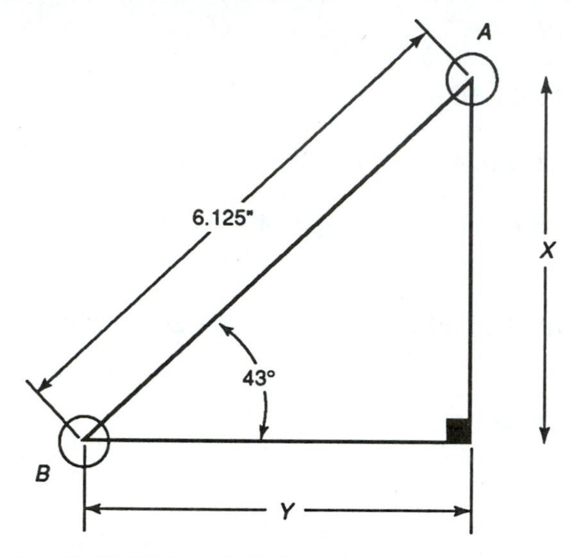

	ANGLE	SIDE *AB*	DIMENSION *X*	DIMENSION *Y*
a.	$\angle B = 43°$	6.125 in.		
b.	$\angle B = 56°$	355 mm		
c.	$\angle B = 2° 41'$	14 in.		
d.	$\angle B = 89° 1'$	190.5 mm		
e.	$\angle A = 72°$	$12\frac{1}{4}$ in.		
f.	$\angle A = 27°$	105 mm		

8. Six holes are equally spaced around a 50-mm diameter circle. Compete this chart of values for the circle.

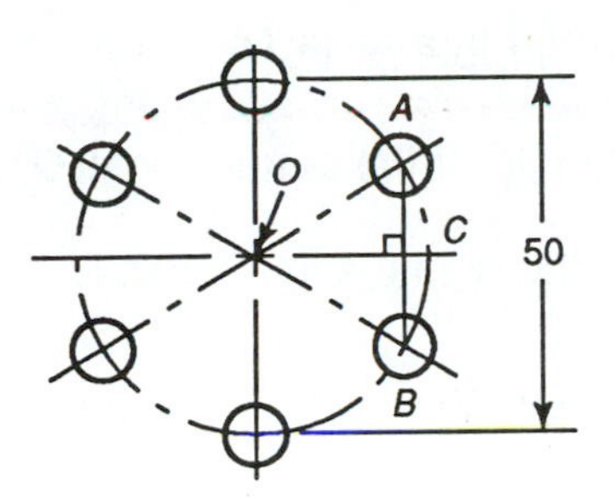

∠ AOB	∠ AOC	$\overline{AO}$	$\overline{AC}$	$\overline{AB}$
a.	b.	c.	d.	e.

9. In order to check the accuracy of the work after boring the holes, the distance across alternate holes at points **A** and **D** is measured. What is the measurement of **AD**? __________

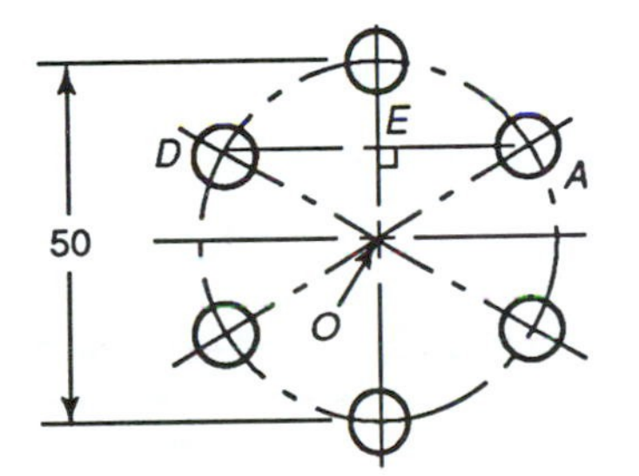

10. Triangle **BAD** is not a right triangle; it is an isosceles triangle. To find the value of angle **BAD**, the perpendicular bisector **AC** is drawn, Side **AC** is 300 mm, side **BD** is 100 mm, and side **BC** is equal to side **CD**. Complete the chart for these values.

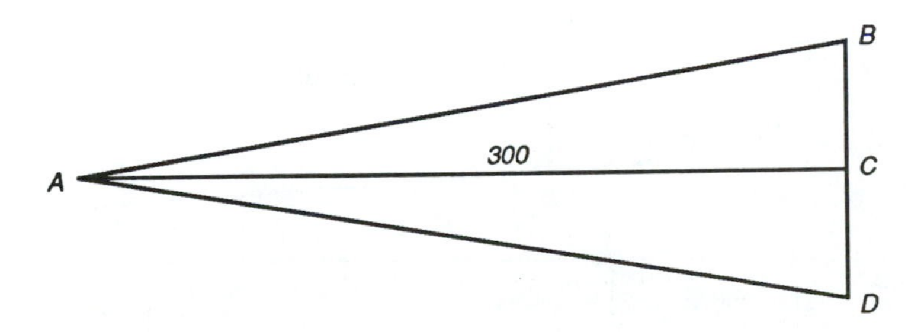

SIDE BC	SIDE AC	∠ BAC (nearer second)	∠ BAD (nearer second)	SIDE AB (nearer whole number)
a.	b.	c.	d.	e.

Note: Use this information for problems 11–14.

To find the included angle of the taper, angle ***O***, line segment ***AC*** is drawn parallel to the centerline. This forms the right triangle ***ABC***. Distance ***BC*** is $\frac{(BD - AE)}{2}$.

The included angle (angle ***O***) is twice the measure of angle ***BAC***.

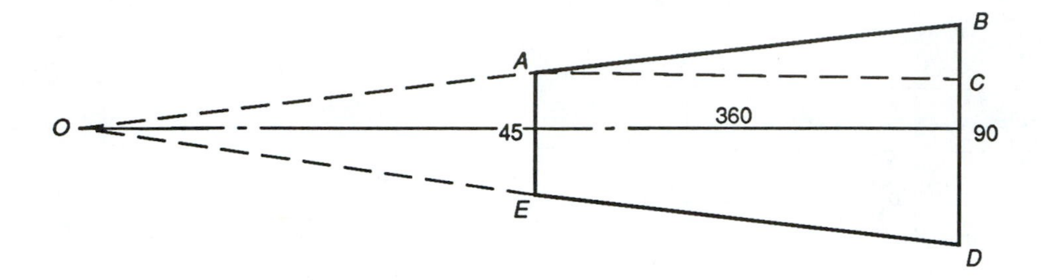

11. In this taper, ***BD*** is 90 millimeters, ***AE*** is 45 millimeters, and the centerline is 360 millimeters. Complete this chart of values.

BC	***∠ BAC* (nearer second)**	***∠ O* (nearer second)**	***AB* (nearer tenth)**
a.	b.	c.	d.

12. The dimensions of different tapers are given in the chart. Complete the chart for the remaining values.

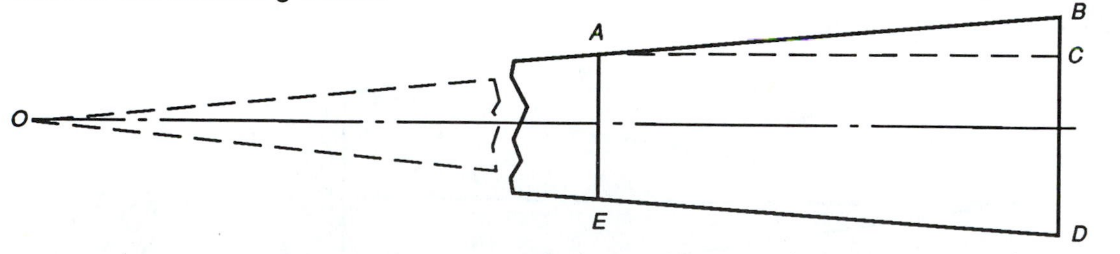

	$\overline{AE}$	$\overline{BD}$	$\overline{AC}$	$\overline{BC}$	∠ ***BAC*** (nearer second)	∠ ***O*** (nearer second)
a.	1¼ in.	2½ in.	5 in.			
b.	79.37 mm	82.55 mm	101.6 mm			
c.	31.75 mm	48.62 mm	57.15 mm			

13. Complete the chart of values for this taper. Express the lengths to the nearer ten-thousandth inch.

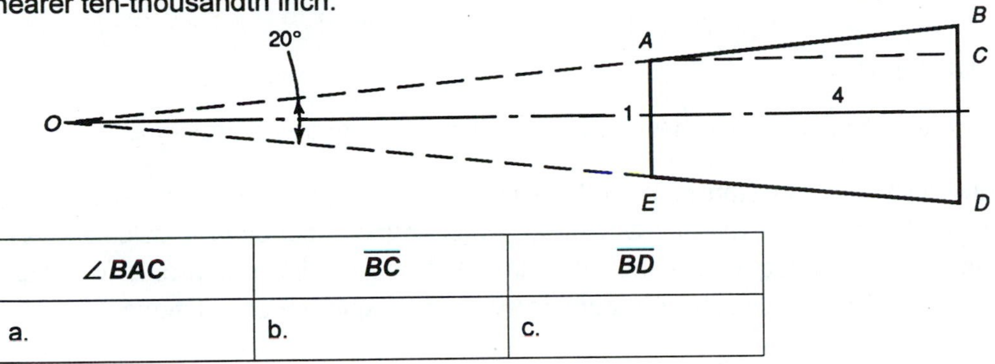

∠ BAC	$\overline{BC}$	$\overline{BD}$
a.	b.	c.

14. Complete the chart of values for this taper. Express the lengths to the nearer ten-thousandth inch.

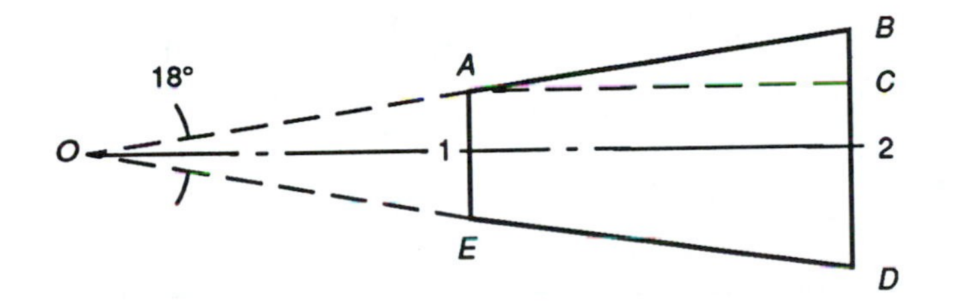

∠ BAC	$\overline{BC}$	$\overline{AC}$	$\overline{AB}$
a.	b.	c.	d.

Unit 45 SINE BAR CALCULATIONS

BASIC PRINCIPLES

- Study and apply the principles of *sine bar calculations* to these problems.

  Since bars are precision tools that are used with gauge blocks to measure, check, or construct angles with extreme accuracy. While normal trigonometric calculations can be used to find the required values, formulas help solve sine bar problems quickly and accurately. These principles are commonly used for sine bar calculations.

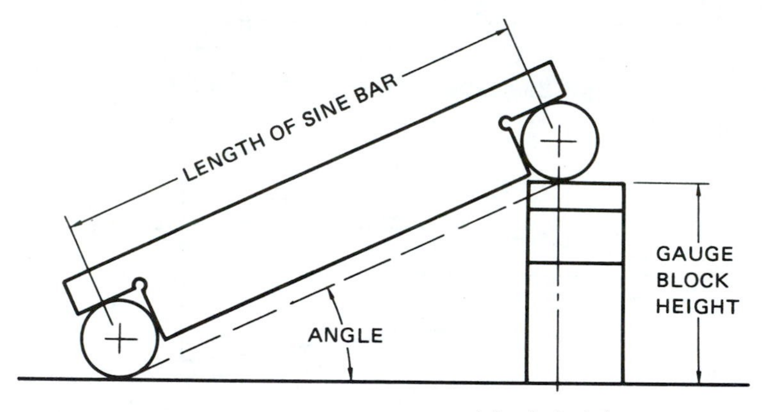

$$\text{sine of angle} = \frac{\text{gauge block height}}{\text{length of sine bar}}$$

$$\text{gauge block stack height} = \text{sine of angle} \times \text{length of sine bar}$$

To find the sine of the angle, the table of trigonometric functions is used.

PRACTICAL PROBLEMS

1. A machinist is required to check this workpiece, which has an angle of 15 degrees 24 minutes. What is the height of the gauge blocks that must be used with a 5-inch sine bar? Express the answer to four decimal places. __________

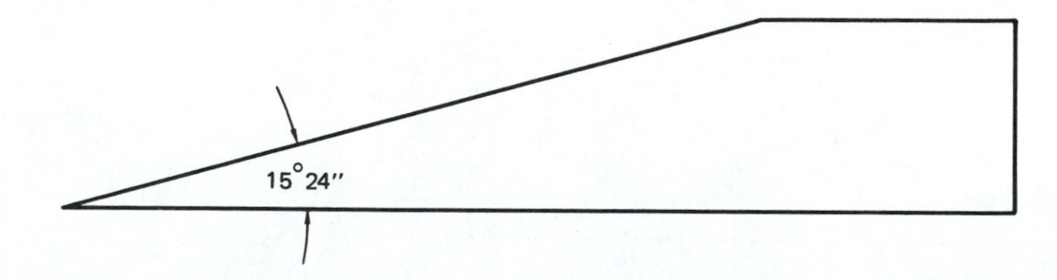

2. A 10-inch sine bar is set to a height of 1.4781 inches. To what angle is the sine bar set? __________

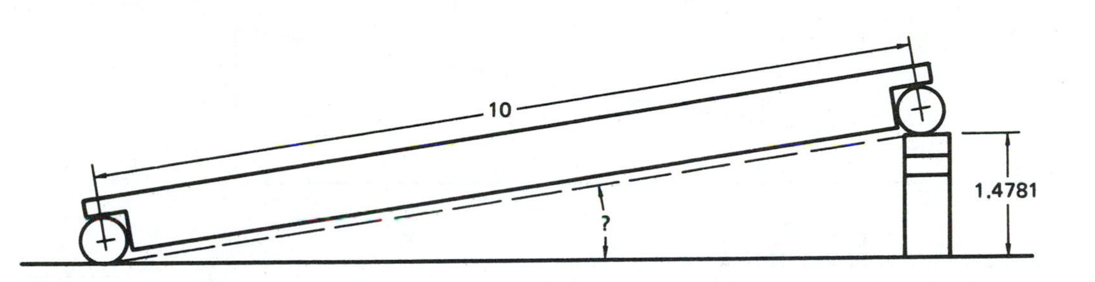

3. Using a 125-mm sine bar, a worker must construct an angle of 30 degrees 15 minutes 30 seconds. How high must the gauge blocks be stacked for this angle? Round the answer to the nearer thousandth millimeter. __________

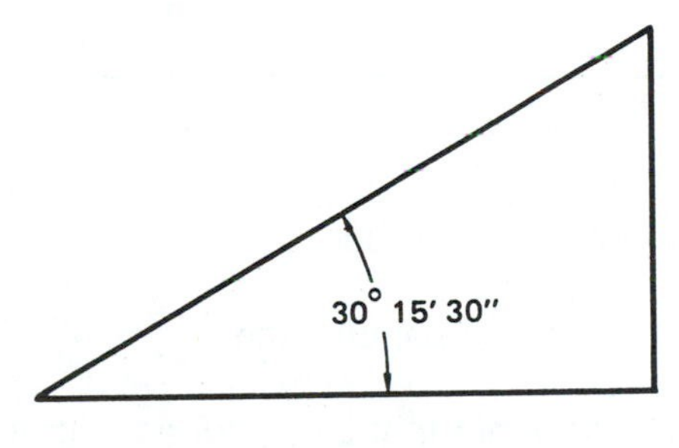

4. This angle is constructed using a stack height of 101.88 mm and a 250-mm sine bar. Find the measure of the angle to the nearer second. __________

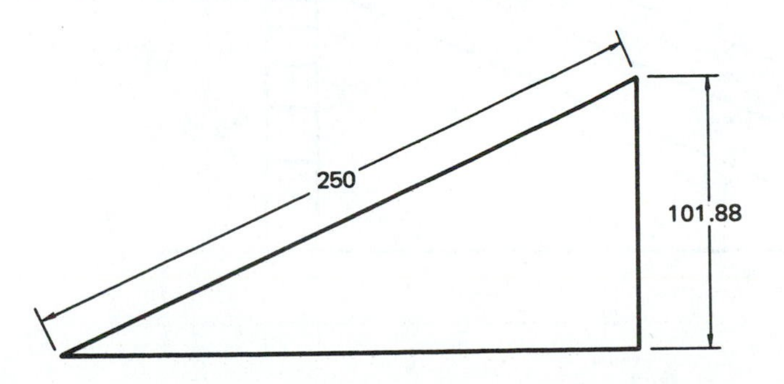

5. This workpiece is checked using a 250-mm sine bar.

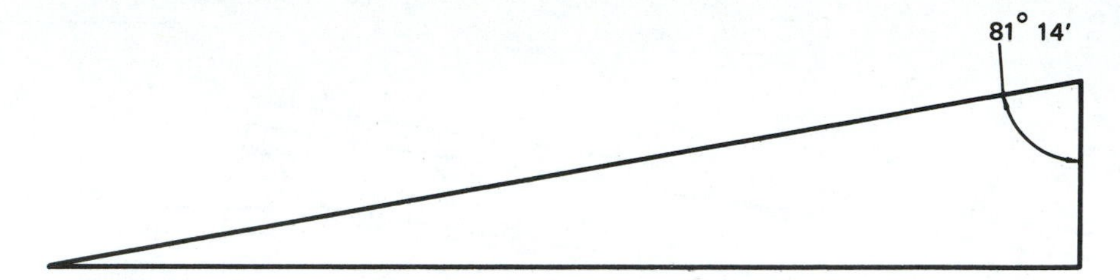

   a. Find the angle of the part. a.__________

   b. Find, to the nearer thousandth millimeter, the stack height required. b.__________

6. This part is checked with a 5-inch sine bar.

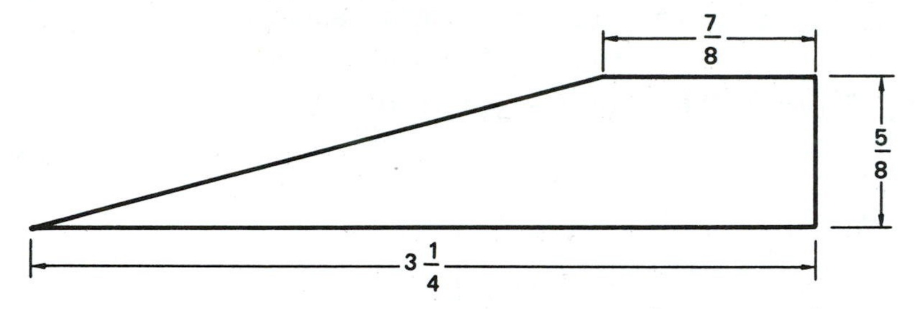

   a. Find the measure of the angle to the nearer second. a.__________

   b. Find the stack height to the nearer ten-thousandth inch. b.__________

7. Find, to the nearer second, the sine bar angle that is produced with the illustrated set-up. The dimensions are in millimeters. __________

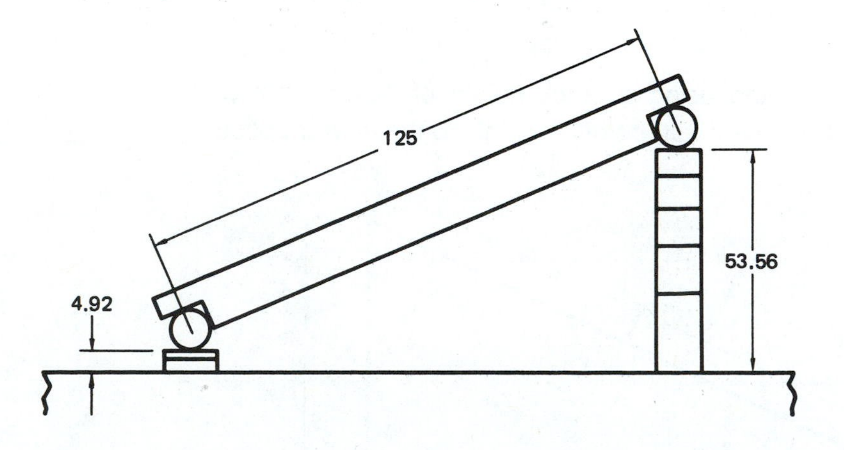

Unit 46 MEASURING ANGLES WITH DISCS

BASIC PRINCIPLES

- Study and apply these principles of *measuring angles with discs* to these problems.

  To measure dovetails, accurately sized steel cylindrical rods or wires are used. These discs are held against the sides of the *V* slots and are accurately measured with a micrometer caliper. To measure with a micrometer caliper is referred to as "to mike." Measuring dovetails is referred to as "to mike over a pair of plugs."

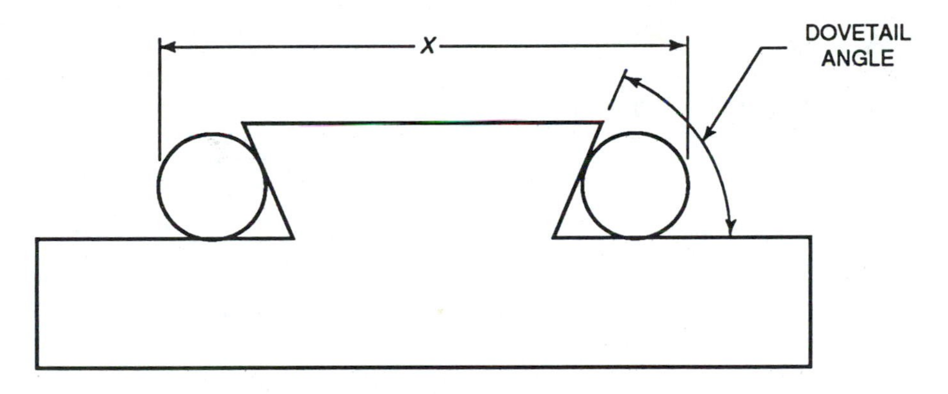

MALE

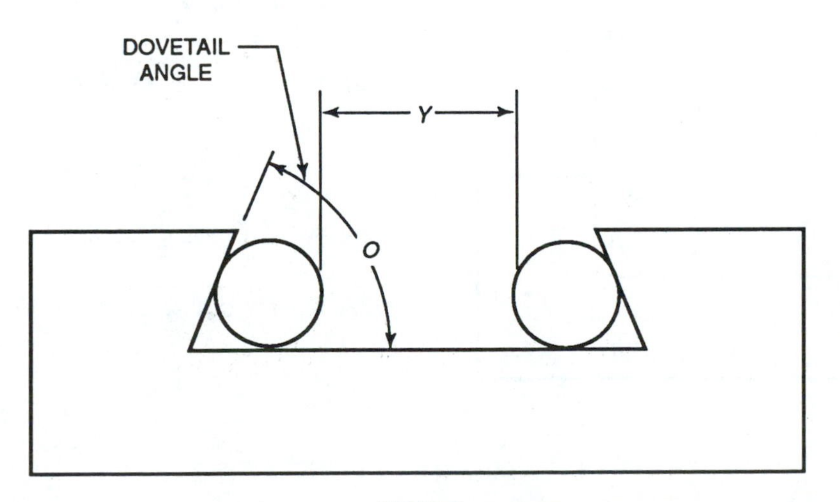

FEMALE

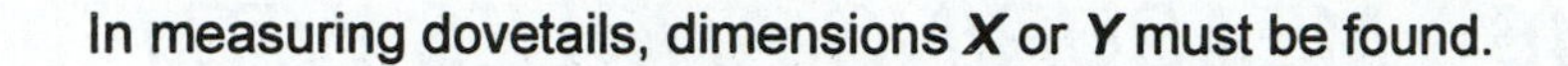
In measuring dovetails, dimensions ***X*** or ***Y*** must be found.

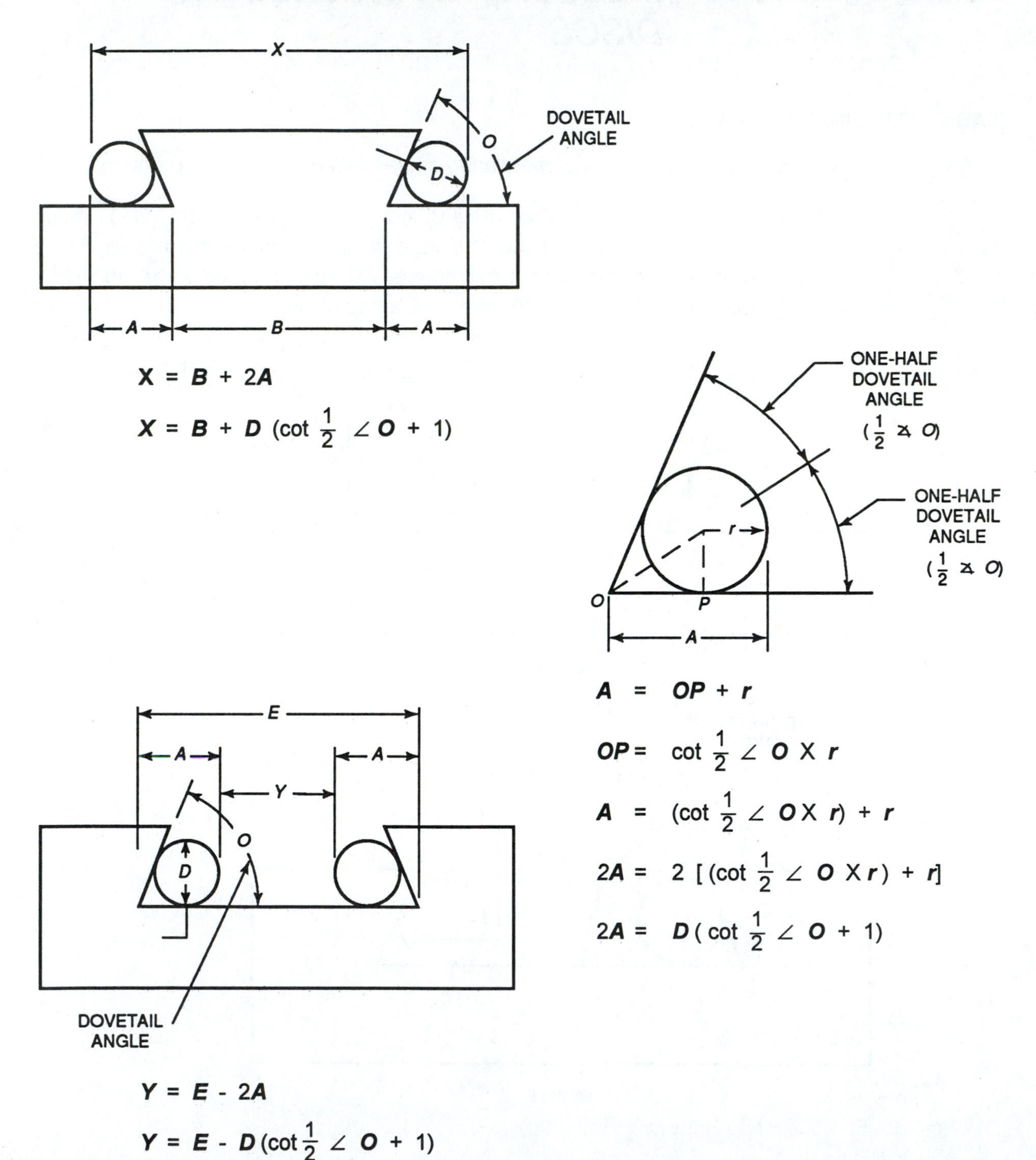

PRACTICAL PROBLEMS

1. This male dovetail is to be measured.

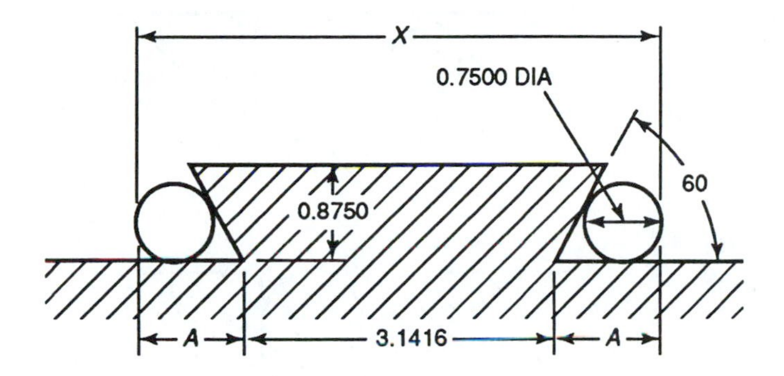

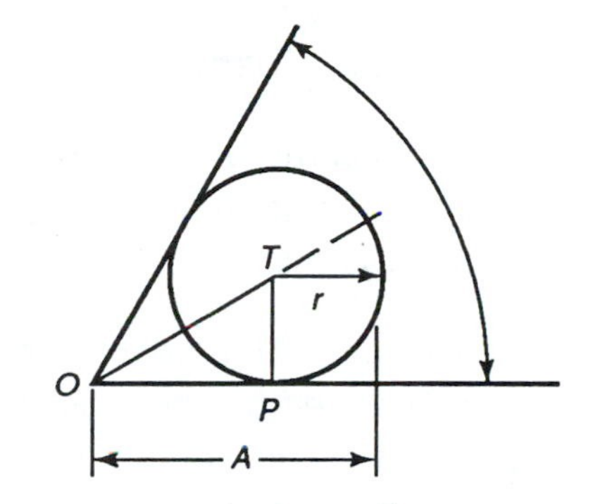

a. What is the value of ½ angle ***O***? a. __________

b. What is the value of the radius? b. __________

c. Find the length of ***OP*** to the nearer ten-thousandth inch. c. __________

d. Find dimension ***A*** to the nearer ten-thousandth inch. d. __________

e. What is the length over the plugs (dimension ***X***) to the nearer ten-thousandth inch? e. __________

Note: Use this diagram for problems 2 and 3.

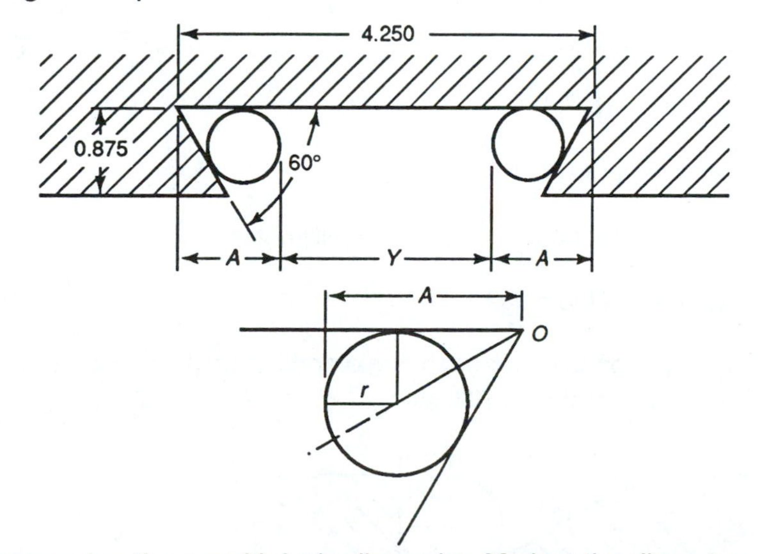

2. Find, to the nearer ten-thousandth inch, dimension ***Y*** when the diameter equals 0.5 inch. __________

3. Find, to the nearer ten-thousandth inch, dimension ***Y*** when the diameter equals $^3/_8$ inch. ________

Note: Use this information for problems 4 and 5.

To accurately measure the angle of a cut, a plug that is flush with the top surface is used.

The size of the needed plug is calculated and turned to size. It is then used to test the accuracy of the cut. The radius or diameter of the plug determines its size. To find the radius this trigonometric function can be used:

$$\tan A = \frac{r}{L - r},$$

where ***A*** is one-half the angle of the cut. The equation is then solved for ***r***.

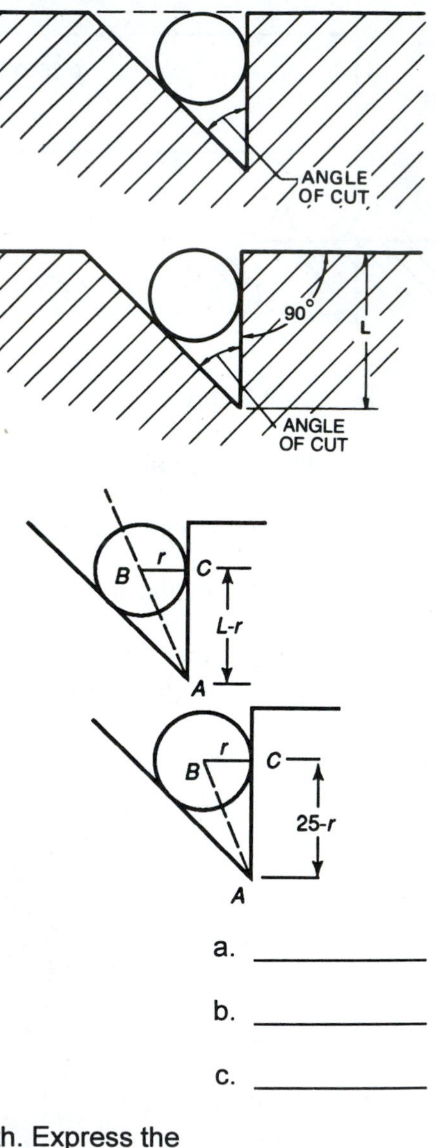

4. This cut is to be measured for accuracy.

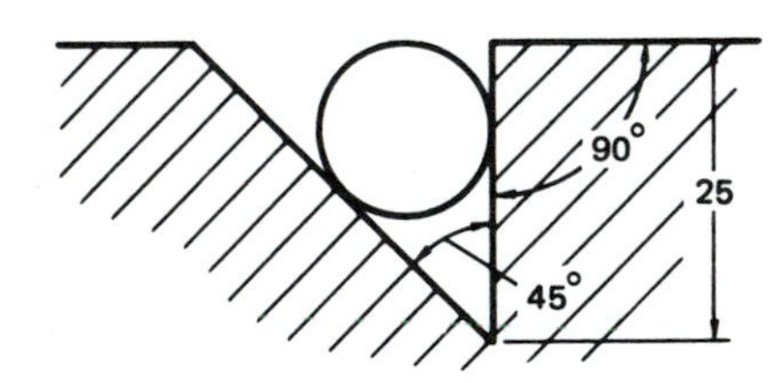

a. What is the value of angle ***A***? a. ________

b. Find the value of ***r*** to the nearer thousandth millimeter. b. ________

c. What is the diameter of the plug? c. ________

5. Find the size of a plug to be used for each angle and length. Express the size of the plug to the nearer tenth millimeter.

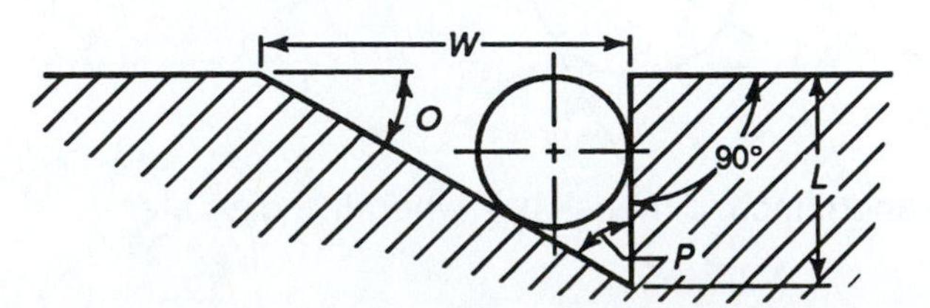

	DIMENSION L	ANGLE O	ANGLE P	SIZE OF PLUG
a.	50 mm	30°		
b.	27 mm	60°		
c.	19.4 mm	24°		

Note: Use this information for problems 6–9.

When great accuracy is needed in measuring angles or in making tapers, discs are used. The discs may either be in contact with each other or a certain distance apart. These formulas are used to find angles, distances and tapers.

DISCS NOT IN CONTACT

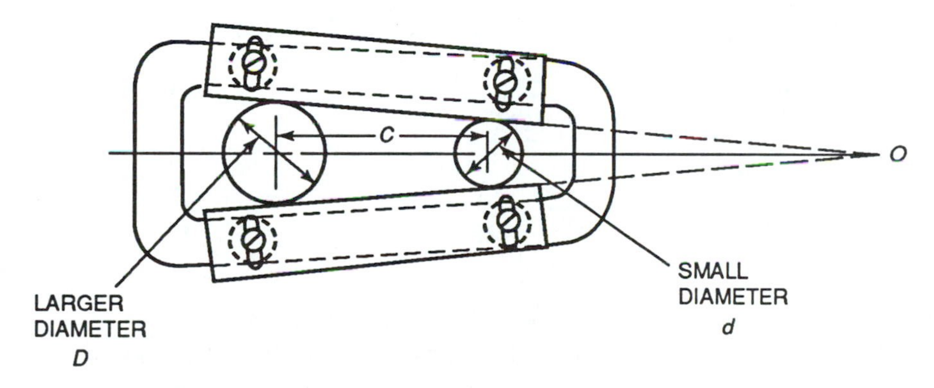

DISCS IN CONTACT

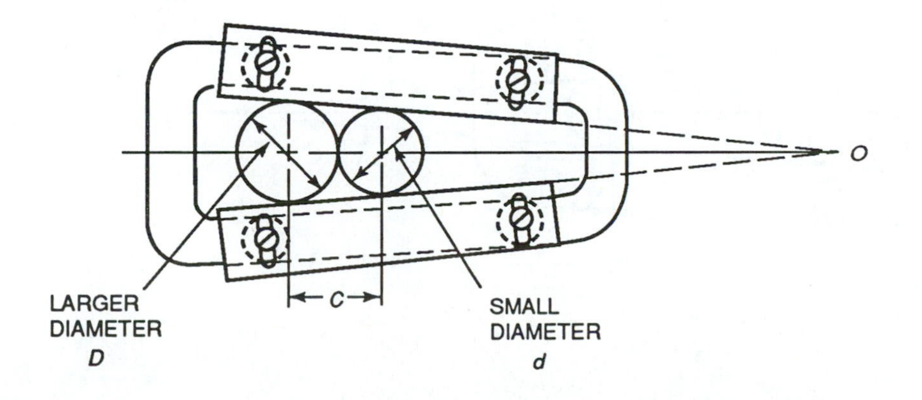

$$\text{taper per inch (millimeter)} = \frac{D - d}{C}$$

$$\tan \frac{1}{2} \angle O = \frac{\text{taper per foot}}{24}$$

$$\sin \frac{1}{2} \angle O = \frac{D - d}{2C}$$

6. Two discs are in contact with each other. Straight edges are placed tangent to the discs. The diameters of the discs are 1.750 inches and 1.250 inches. Find, to the nearer second, angle ***O*** ____________

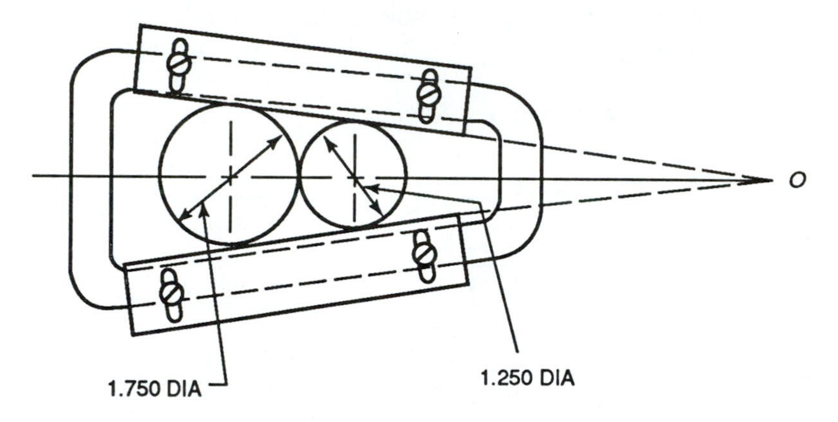

7. Two standard reference discs are placed to lay out an angle of 9 degrees 27 minutes. The diameters of the discs are 25 millimeters and 50 millimeters. Find, to the nearer hundredth millimeter, the center distance ***C***. ____________

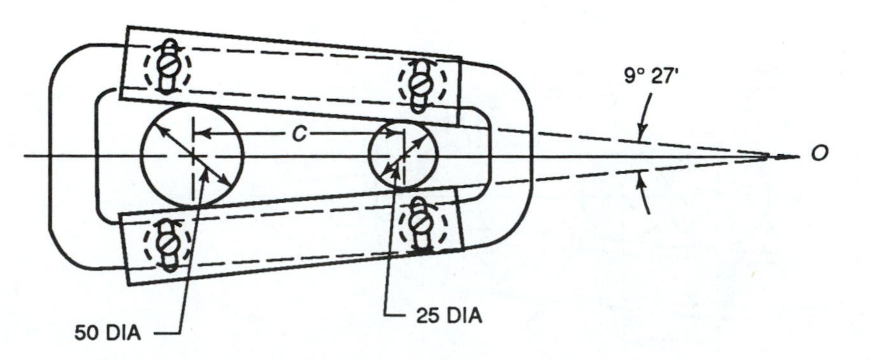

8. Two discs, 22.23 mm and 57.15 mm diameter, are used to measure a taper. The distance between the centers is 98.43 mm. What is the rate of taper per millimeter? Round the answer to four decimal places. ____________

9. A piece of work tapers at a rate of $\frac{7}{8}$ inch per foot. The included angle of the taper is measured by two discs with diameters of $\frac{9}{16}$ inch and $1\frac{3}{16}$ inches.

   a. Find the included angle to the nearer second. a. ____________

   b. Find the distance between the centers to the nearer ten-thousandth inch. b. ____________

Unit 47 HELIX ANGLES

BASIC PRINCIPLES

- Study and apply these principles of *helix angles* to these problems.

  In cutting a thread on the surface of a cylinder, the point of the tool makes a curved line. This curved line is a *helix*. The horizontal distance advanced by the helix in making a complete revolution is called the *lead* of the screw.

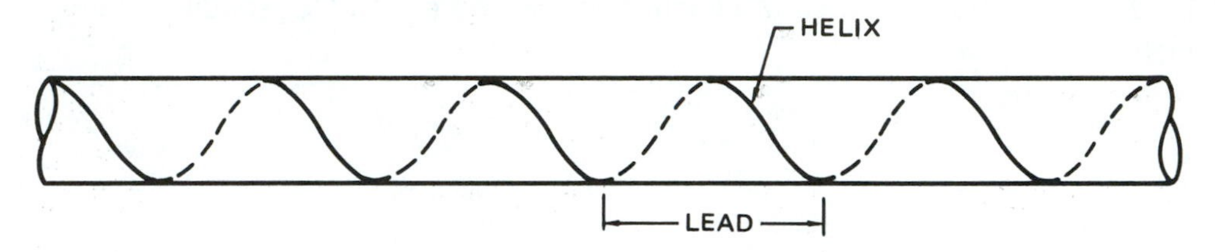

Suppose one turn of the helix could be unwrapped from the surface of the cylinder. The helix becomes a straight line forming the hypotenuse **AB** of the right triangle **ABC**. Side **AC** would be equal to the lead. Side **BC** would be equal to the circumference of the cylinder. Angle **BAC** is the helix angle for drills, cutters, and gears. Angle **ABC** is the helix angle for screw threads and worm threads. This angle is often called the *lead angle*.

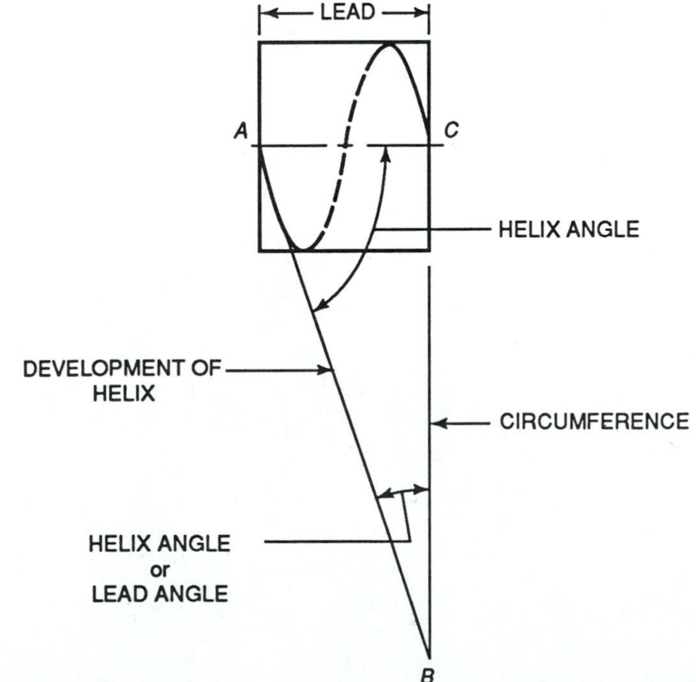

PRACTICAL PROBLEMS

Note: Use this information for problems 1-3.

The curved line of screw threads is one example of a helical curve. In cutting screw threads on a lathe, the tool travels along and parallel to the centerline of the work and the work rotates. The tool cuts a helical groove in the work.

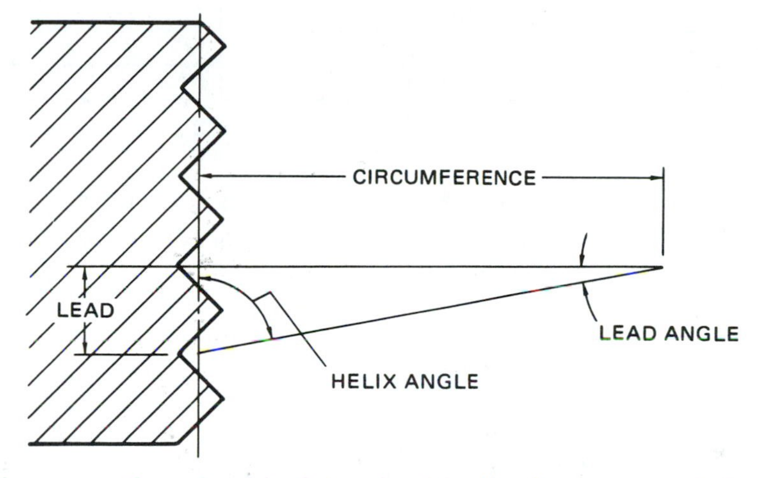

The lead of the screw thread cut is determined by the feed per revolution of the work. A combination of gears connecting the spindle and lead screw of the lathe produce the correct amount of feed per revolution of the work.

The lead angle is found by using trigonometric formulas. The lead equals the distance that a thread advances in one turn, or the pitch.

$$\text{tangent of lead angle} = \frac{\text{lead}}{\text{circumference}}$$

$$\text{cotangent of lead angle} = \frac{3.1416 \times \text{pitch diameter of thread}}{\text{lead of screw thread}}$$

1. Find the lead angle of each screw thread. Round each answer to the nearer minute.

	SIZE	PITCH DIAMETER	LEAD ANGLE
a.	1/4 – 20	0.2141 in.	
b.	5/16 – 18	0.2751 in.	
c.	M10 × 1.5	9.13 mm	
d.	M12 × 1.75	10.95 mm	
e.	1/2 – 20	0.4662 in.	
f.	3/4 – 10	0.6832 in.	
g.	M16 × 2	14.81 mm	
h.	M24 × 3	22.23 mm	

2. In order to grind the proper clearance on a threading tool, it is necessary to know the lead angle of the thread. Find the lead angle of each *ACME* thread. Express each answer to the nearer minute.

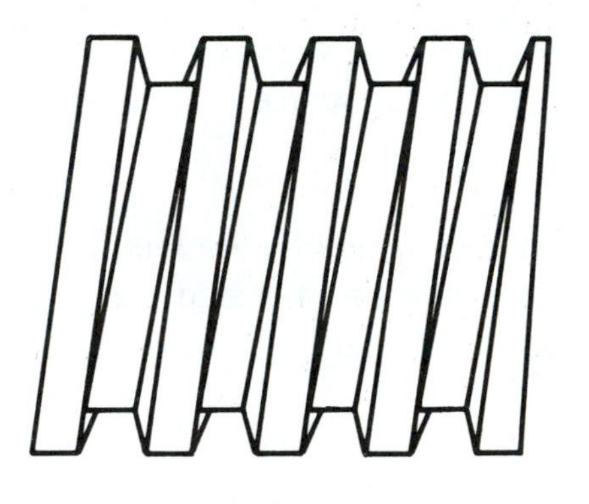

	SIZE	PITCH DIAMETER	LEAD ANGLE
a.	$\frac{5}{16}$ - 14	0.2728 in.	
b.	$\frac{5}{8}$ - 8	0.5562 in.	
c.	1 - 5	0.8759 in.	
d.	$1\frac{3}{8}$ - 4	1.2214 in.	

3. Find the lead angle for each metric trapezoid thread. Round each answer to the nearer minute.

	SIZE	PITCH DIAMETER	LEAD ANGLE
a.	10 mm X 3 mm	8.5 mm	
b.	22 mm X 5 mm	19.5 mm	
c.	36 mm X 6 mm	33 mm	
d.	48 mm X 8 mm	44 mm	

Note: Use this information for problems 4–7.

For drills, cutters, and gears, the helix angle is found by using trigonometric formulas.

$$\textit{tangent of helix angle} = \frac{\textit{circumference}}{\textit{lead}}$$

or

$$\text{tangent of helix angle} = \frac{3.1416 \times \textit{diameter}}{\textit{lead}}$$

$$\textit{lead} = \frac{3.1416 \times \textit{diameter}}{\textit{tangent of helix angle}}$$

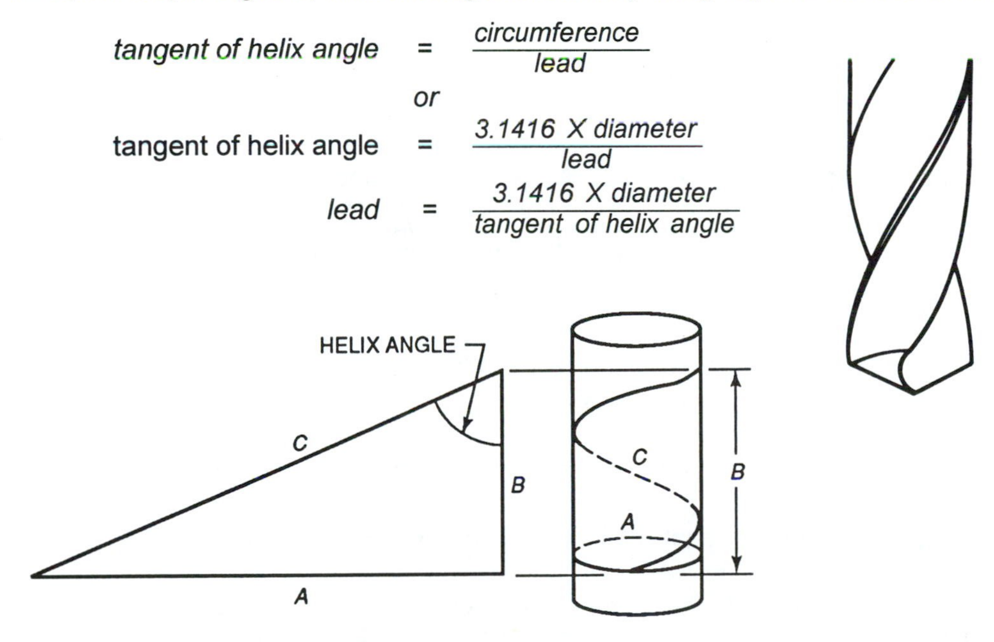

4. Find the helix angle for these values involving flute milling. Round each answer to the nearer minute.

	DIAMETER	LEAD	HELIX ANGLE
a.	$1\frac{1}{2}$ in.	9 in.	
b.	55 mm	404 mm	
c.	$\frac{3}{4}$ in.	6.312 in.	
d.	29.7 mm	99.6 mm	

5. Find the lead of each helical tool to the nearer thousandth.

	TOOL	DIAMETER	HELIX ANGLE	LEAD
a.	Drill	1/2 in.	10°	
b.	Slab Cutter	100 mm	12°	
c.	End Mill	50 mm	25°	

6. Find the helix angle of each helical gear. Express each answer to the nearer minute.

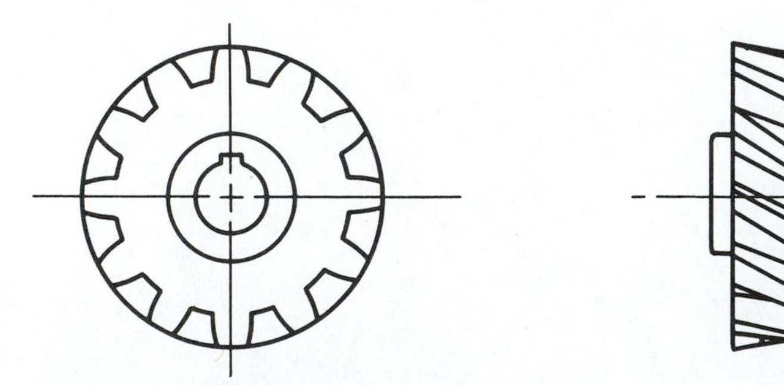

	PITCH DIAMETER	LEAD	HELIX ANGLE
a.	4.695 in.	6.567 in.	
b.	12.543 in.	88.512 in.	
c.	59.7 mm	172.8 mm	
d.	74.3 mm	209.7 mm	
e.	8.531 in.	23.315 in.	

7. Find the lead of each gear to the nearer thousandth.

	PITCH DIAMETER	HELIX ANGLE	LEAD
a.	94.2 mm	27° 30′	
b.	144.8 mm	54°	
c.	2.771 in.	61°	
d.	171.8 mm	43° 29′	

Unit 48 ACUTE TRIANGLES

BASIC PRINCIPLES

- *Acute angles* are those that are less than 90°. The principles of right angle trigonometry learned in Unit 47 may be applied to the solution of the following problems.

PRACTICAL PROBLEMS

1. A machine operator must drill the holes in this lever. To do this operation the machinist must use a vertical milling machine. The table movements are only at right angles. Find each dimension so this part can be drilled on the milling machine. Express each answer to the nearer hundredth millimeter.

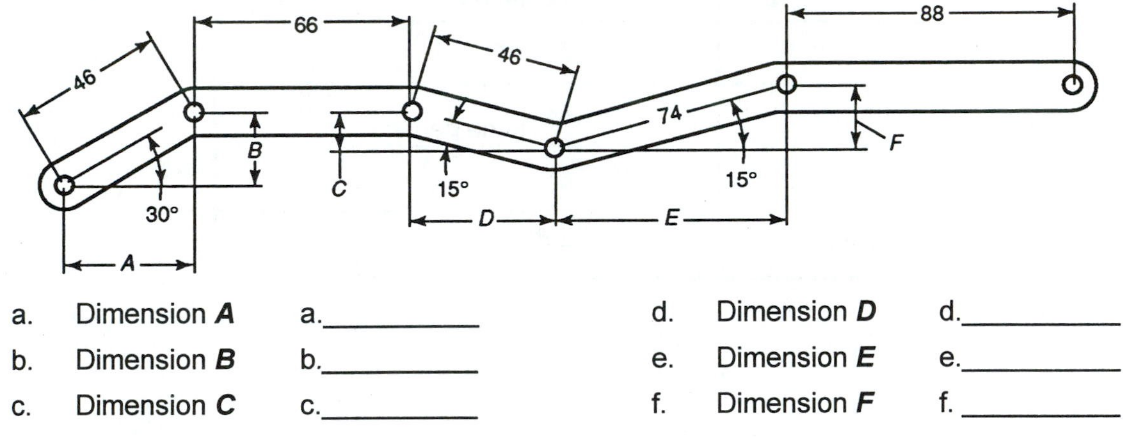

a.	Dimension ***A***	a.__________	d.	Dimension ***D***	d.__________
b.	Dimension ***B***	b.__________	e.	Dimension ***E***	e.__________
c.	Dimension ***C***	c.__________	f.	Dimension ***F***	f. __________

2. For a piercing operation, a diemaker must grind a shear angle on the face of a punch. The stock thickness is 0.063 inch. The desired shear is equal to 1.5 times the stock thickness. Find, to the nearer minute, the angle to be ground on the punch. __________

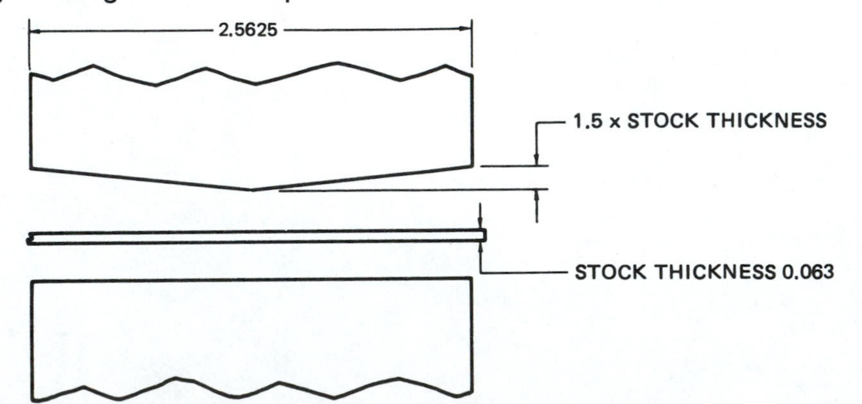

3. When cutting threads on a lathe the operator calculates the proper depth of threads to be 0.0465 inch. This means that the operator must move the tool into the work 0.0465 inch. if a 30 degree tool angle is used, how far must the tool actually be advanced with the compound rest? Express the answer to the near ten-thousandth inch. __________

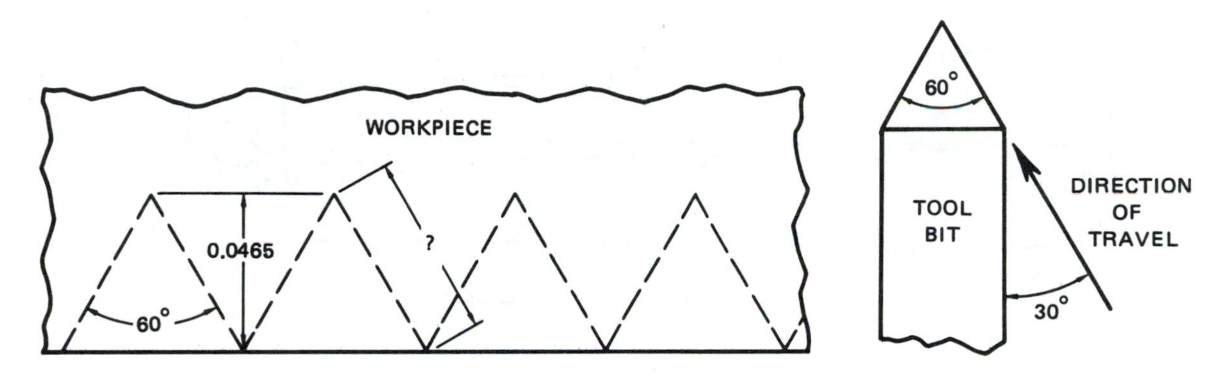

4. A machinist receives a work order to make this lathe center.

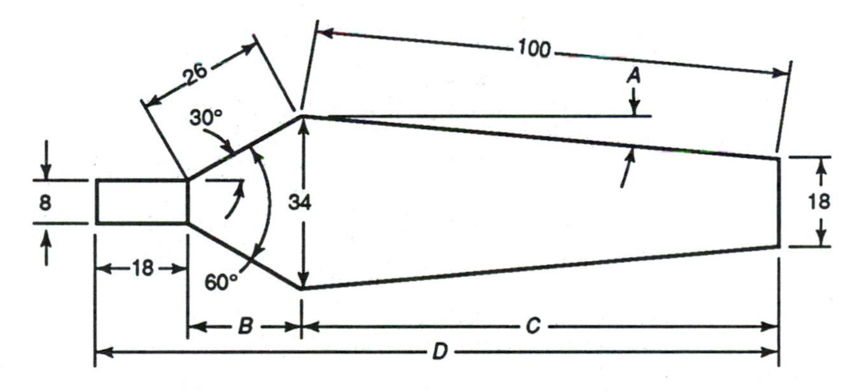

a. Find angle ***A*** to the nearer second. a. __________

b. Find dimension ***B*** to the nearer tenth millimeter. b. __________

c. Find dimension ***C*** to the nearer tenth millimeter. c. __________

d. Find dimension ***D*** to the nearer tenth millimeter. d. __________

5. A common method used to turn tapers on a lathe is by offsetting the tailstock. This part is 12 inches long and has an angle of taper of 1 degree 30 minutes. How much will the tailstock be offset to cut the taper? __________

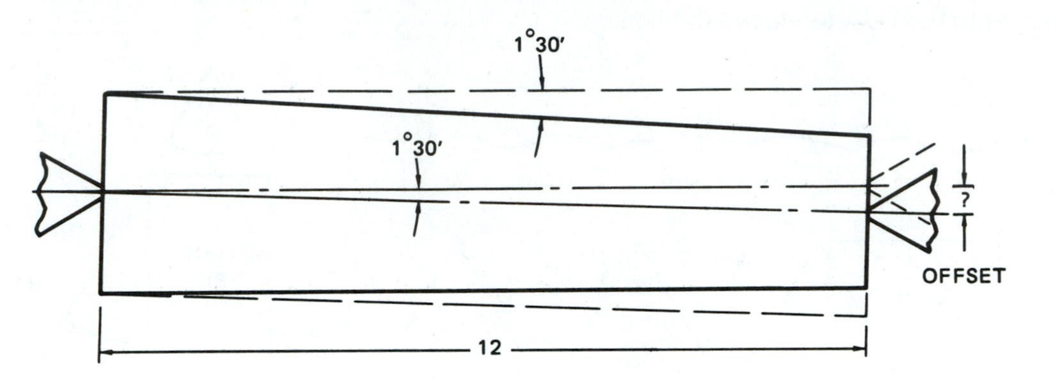

Unit 49 OBLIQUE TRIANGLES

BASIC PRINCIPLES

- Study and apply these principles of *oblique triangles* to these problems.

  To find the value of the function of angles greater than 90°, a reference angle is used. The reference angle is an angle equal to or less than 90°.

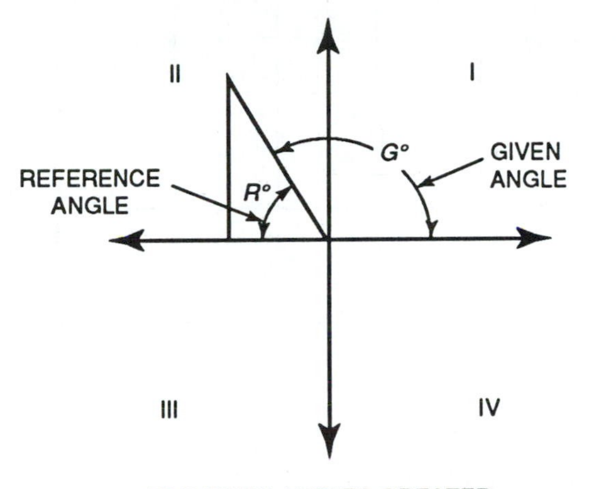

FOR GIVEN ANGLES GREATER THAN 90° AND LESS THAN 180°

$180° - G° = R°$

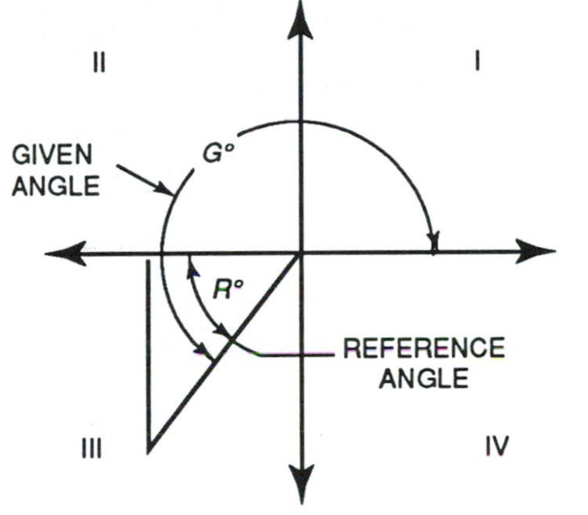

FOR GIVEN ANGLES GREATER THAN 180° AND LESS THAN 270°

$G° - 180° = R°$

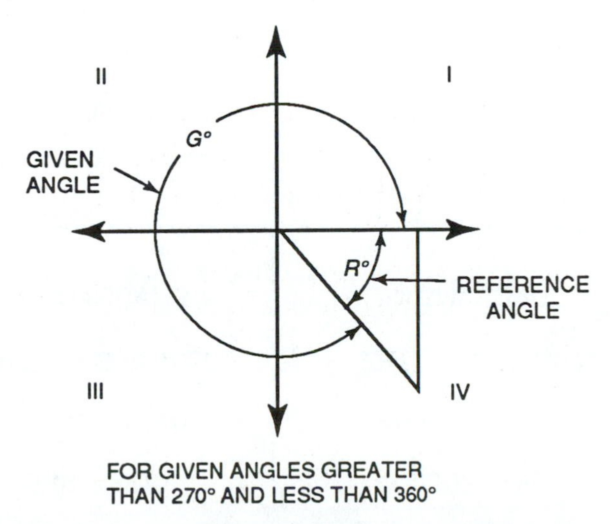

FOR GIVEN ANGLES GREATER THAN 270° AND LESS THAN 360°

$360° - G° = R°$

The sign of the functions depends on the measure of the angle or the quadrant the angle is in.

This chart shows the sign of the functions for each quadrant.

QUADRANT	MEASURES OF ANGLES	SIN	COS	TAN	CSC	SEC	COT
I	Less than 90°	+	+	+	+	+	+
II	Greater than 90° less than 180°	+	-	-	+	-	-
III	Greater than 180° less than 270°	-	-	+	-	-	+
IV	Greater than 270° less than 360°	-	+	-	-	+	-

This diagram summarizes the values of the functions for angles.

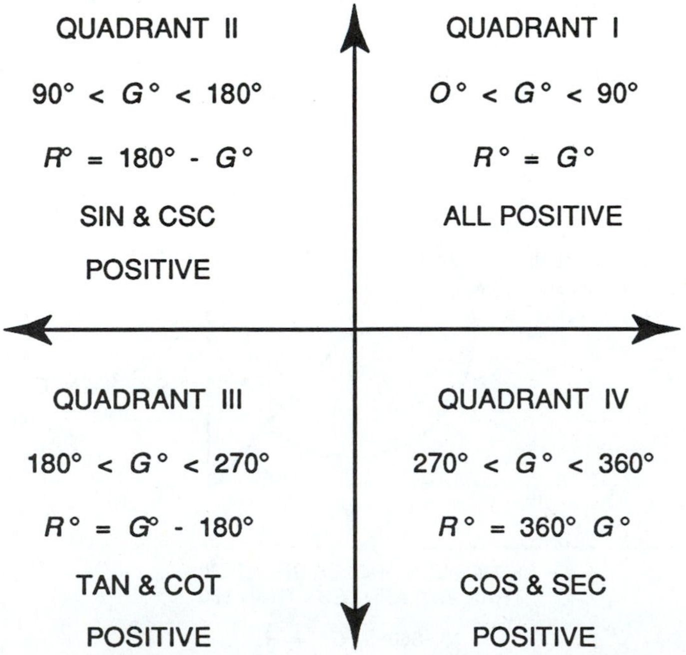

QUADRANT II

$90° < G° < 180°$

$R° = 180° - G°$

SIN & CSC POSITIVE

QUADRANT I

$O° < G° < 90°$

$R° = G°$

ALL POSITIVE

QUADRANT III

$180° < G° < 270°$

$R° = G° - 180°$

TAN & COT POSITIVE

QUADRANT IV

$270° < G° < 360°$

$R° = 360° \; G°$

COS & SEC POSITIVE

PRACTICAL PROBLEMS

1. Complete this chart for the value of each function.

	ANGLE	SINE	COSINE	TANGENT
a.	127°			
b.	149° 18′			
c.	218° 41′			
d.	98° 21′ 47″			
e.	328° 42′ 51″			
f.	142° 7′ 42″			
g.	298° 41′ 7″			
h.	171° 14′ 27″			
i.	164° 18′ 51″			

2. For each function, find two values of the angle to the nearer second. One value will be an angle in the first quadrant. One other value will be in either the second, third or fourth quadrant.

	FUNCTION	VALUE	QUADRANT I ANGLE	OTHER QUADRANT ANGLE
a.	sin	0.15931		
b.	cot	0.31690		
c.	cos	0.91793		
d.	tan	0.37143		

Note: Use this information for problems 3 and 4.

Dimensions on oblique triangles are found by drawing perpendiculars to make right triangles and using ratio. Two ratios that are commonly used are the *Law of Sines* and the *Law of Cosines*.

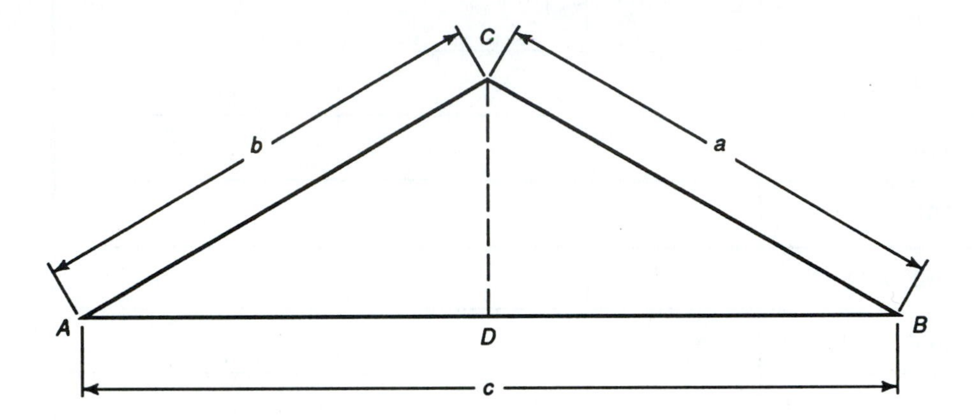

LAW OF SINES

$$\frac{\sin A}{\sin B} = \frac{a}{b}$$

$$\frac{\sin A}{\sin C} = \frac{a}{c}$$

$$\frac{\sin B}{\sin C} = \frac{b}{c}$$

LAW OF COSINES

$$\cos A = \frac{b^2 + c^2 - a^2}{2bc}$$

$$a = \sqrt{b^2 + c^2 - 2bc \cos A}$$

$$\cos B = \frac{a^2 + c^2 - b^2}{2ac}$$

$$b = \sqrt{a^2 + c^2 - 2ac \cos B}$$

$$\cos C = \frac{a^2 + b^2 - c^2}{2ab}$$

$$c = \sqrt{a^2 + b^2 - 2ab \cos C}$$

3. Using the Law of Sines or the Law of Cosines, complete this chart. Express each side to the nearer thousandth.

	SIDE *a*	SIDE ***b***	**SIDE *c***	**ANGLE *A***	ANGLE ***B***	**ANGLE *C***
a.		91 mm		46° 19′	71° 39′	
b.			21.367 mm	17° 17′ 41″		130° 42é 18″
c.	42.3 mm			67° 18′ 14″	48° 13é 8″	
d.		2,674 mm		32° 14′		92° 19é

4. A toolmaker is required to lay out three holes in a plate. Find, to the nearer hundredth millimeter, side ***b*** of the triangle that is formed. __________

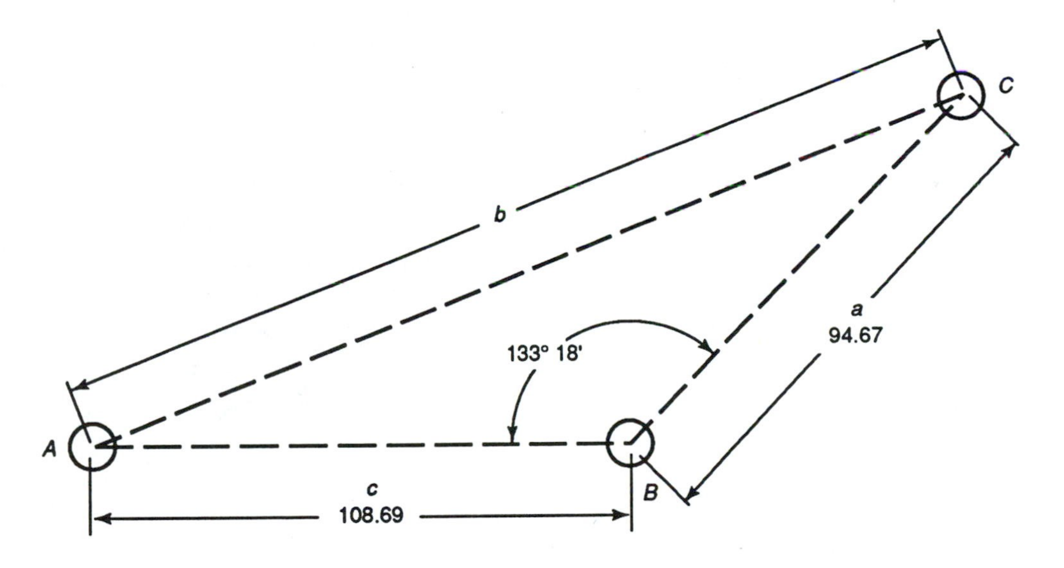

CRITICAL THINKING PROBLEMS

1. Normally countersinks are dimensioned by calling out the diameter and the angle as shown below.

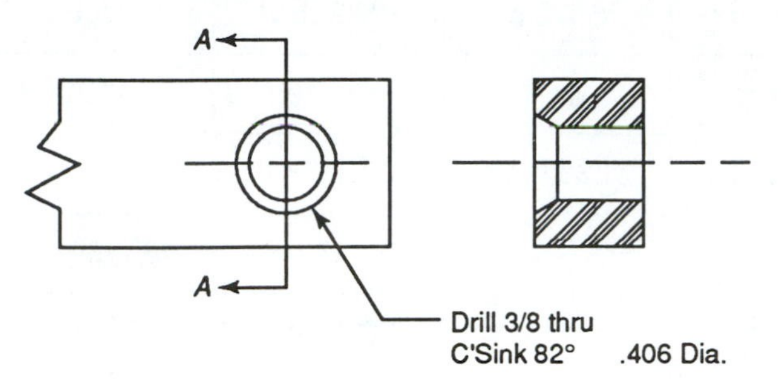

Determine the depth that the countersink must go to obtain a diameter as shown. __________

2.

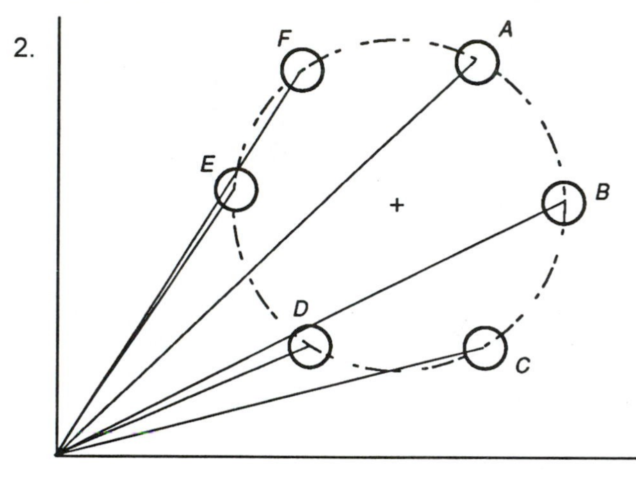

Determine the *x*, *y* coordinates of the holes shown, relative to the REF point. Polar coordinates are given.

POLAR COORDINATES

A 3.4421″ @ 43.4229°
B 3.3541″ @ 26.5651°
C 2.5791″ @ 14.2297°
D 1.6285″ @ 22.9113°
E 1.8028″ @ 56.3099°
F 2.8014″ @ 57.6263°

	X	**Y**
A		
B		
C		
D		
E		
F		

3. Referring to problem #2 above, determine by inspection the radius of the bolt circle and the *x*, *y* coordinates of the bolt circle center point relative to the REF point. ____________

APPENDIX

SECTION 1

DENOMINATE NUMBERS

Denominate numbers are numbers that include units of measurement. The units of measurement are arranged from the largest units at the left to the smallest unit at the right.

For example: 6 yd 2 ft 4 in.

All basic operations of arithmetic can be performed on denominate numbers.

I. EQUIVALENT MEASURES

Measurements that are equal can be expressed in different terms. For example, 12 in = 1 ft. If these equivalents are divided, the answer is 1.

$$\frac{1 \text{ ft}}{12 \text{ in}} = 1 \qquad \frac{12 \text{ in}}{1 \text{ ft}} = 1$$

To express one measurement as another equal measurement, multiply by the equivalent in the form of 1.

To express 6 inches in equivalent foot measurement, multiply 6 inches by one in the form of $\frac{1 \text{ ft}}{12 \text{ in}}$. In the numerator and denominator, divide by a common factor.

$$6 \text{ in} = \frac{\overset{1}{\cancel{6 \text{ in}}}}{1} \times \frac{1 \text{ ft}}{\underset{2}{\cancel{12 \text{ in}}}} = \frac{1}{2} \text{ ft or } 0.5 \text{ ft.}$$

To express 4 feet in equivalent inch measurement, multiply 4 feet by one in the form of $\frac{12 \text{ in}}{1 \text{ ft}}$.

$$4 \text{ ft} = \overset{4}{\cancel{4 \text{ ft}}} \times \frac{12 \text{ in}}{\underset{1}{\cancel{1 \text{ ft}}}} = \frac{48 \text{ in}}{1} = 48 \text{ in}$$

Per means division, as with a fraction bar. For example, 50 miles per hour can be written $\frac{50 \text{ miles}}{1 \text{ hour}}$.

II. BASIC OPERATIONS

A. ADDITION

SAMPLE: 2 yd 1 ft 5 in + 1 ft 8 in + 5 yd 2 ft

1. Write the denominate numbers in a column with like units in the same column.

	2 yd	1 ft	5 in
		1 ft	8 in
+	5 yd	2 ft	

2. Add the denominate numbers in each column.

7 yd	4 ft	13 in

3. Express the answer using the largest possible units.

7 yd			= 7 yd	
4 ft			= 1 yd 1 ft	
13 in			=+	1 ft 1 in
7 yd	4 ft	13 in	= 8 yd	2 ft 1 in

B. SUBTRACTION

SAMPLE: 4 yd 3 ft 5 in - 2 yd 1 ft 7 in

1. Write the denominate numbers in columns with like units in the same column.

	4 yd	3 ft	5 in
–	2 yd	1 ft	7 in

2. Starting at the right, examine each column to compare the numbers. If the bottom number is larger, exchange one unit from the column at the left for its equivalent. Combine like units.

7 in is larger than 5 in

3 ft = 2 ft 12 in

12 in + 5 in = 17 in

3. Subtract the denominate numbers.

	4 yd	2 ft	17 in
–	2 yd	1 ft	7 in
	2 yd	1 ft	10 in

4. Express the answer using the largest possible units.

2 yd	1 ft	10 in

C. MULTIPLICATION

– By a constant

SAMPLE: 1 hr 24 min X 3

1. Multiply the denominate number by the constant.

1 hr	24 min
	x 3
3 hr	72 min

2. Express the answer using the largest possible units.

3 hr			
	72 min	= 3 hr	
		= 1 hr	12 min
3 hr	72 min	= 4 hr	12 min

– By a denominate number expressing linear measurement

SAMPLE: 9 ft 6 in X 10 ft

1. Express all denominate numbers in the same unit.

$9 \text{ ft } 6 \text{ in} = \quad 9\frac{1}{2} \text{ ft}$

2. Multiply the denominate numbers. (This includes the units of measure, such as ft X ft = sq ft).

$9\frac{1}{2} \text{ ft} \times 10 \text{ ft} =$

$\frac{19}{2} \text{ ft} \times 10 \text{ ft} =$

95 sq ft

– By a denominate number expressing square measurement

SAMPLE: 3 ft X 6 sq ft

1. Multiply the denominate numbers. (This includes the units of measure, such as ft X ft = sq ft and sq ft X ft = cu ft).

3 ft X 6 sq ft = 18 cu ft

– By a denominate number expressing rate

SAMPLE: 50 miles per hour X 3 hours

1. Express the rate as a fraction using the fraction bar for *per*.

$\frac{50 \text{ miles}}{1 \text{ hour}} \times \frac{3 \text{ hours}}{1} =$

2. Divide the numerator and denominator by any common factors, including unitsofmeasure.

$\frac{50 \text{ miles}}{\overset{}{\underset{1}{\cancel{1 \text{ hour}}}}} \times \frac{\overset{3}{\cancel{3 \text{ hours}}}}{1} =$

3. Multiply numerators. Multiply denominators. — $\frac{150 \text{ miles}}{1} =$

4. Express the answer in the remaining unit. — 150 miles

D. DIVISION

– By a constant

SAMPLE: 8 gal 3 qt ÷ 5

1. Express all denominate numbers in the same unit. — 8 gal 3 qt = 35 qt

2. Divide the denominate number by the constant. — 35 qt ÷ 5 = 7 qt

3. Express the answer using the largest possible units. — 7 qt = 1 gal 3 qt

– By a denominate number expressing linear measurement

SAMPLE: 11 ft 4 in ÷ 8 in

1. Express all denominate numbers in the same unit. — 11 ft 4 in = 136 in

2. Divide the denominate numbers by a common factor. (This includes the units of measure, such as inches ÷ inches = 1.) — 136 in ÷ 8 in =

$$\frac{\overset{17}{\cancel{136 \text{ in}}}}{\underset{1}{\cancel{8 \text{ in}}}} = \frac{17}{1} = 17$$

– By a linear measure with a square measurement as the dividend

SAMPLE: 20 sq ft ÷ 4 ft

1. Divide the denominate numbers. (This includes the units of measure, such as sq ft ÷ ft = ft) — 20 sq ft ÷ 4 ft

$$\frac{\overset{5 \text{ ft}}{\cancel{20 \text{ sq ft}}}}{\cancel{4 \text{ ft}}} = \frac{5 \text{ ft}}{1}$$

2. Express the answer in the remaining unit. — 5 ft

– By denominate numbers used to find rate

SAMPLE: 200 mi ÷ 10 gal

1. Divide the denominate numbers.

$$\frac{\overset{20 \text{ mi}}{\cancel{200 \text{ mi}}}}{\underset{1 \text{ gal}}{\cancel{10 \text{ gal}}}} = \frac{20 \text{ mi}}{1 \text{ gal}}$$

2. Express the units with the fraction bar meaning *per.*

$$\frac{20 \text{ mi}}{1 \text{ gal}} = 20 \text{ miles per gallon}$$

Note: Alternate methods of performing the basic operations will produce the same result. The choice of method is determined by the individual.

SECTION II

EQUIVALENTS

ENGLISH RELATIONSHIPS

ENGLISH LENGTH MEASURE

1 foot (ft.)	=	12 inches (in.)
1 yard (yd.)	=	3 feet (ft.)
1 mile (mi.)	=	1,760 yards (yd.)
1 mile (mi.)	=	5,280 feet (ft.)

ENGLISH AREA MEASURE

1 square yard (sq. yd.)	=	9 square feet (sq. ft.)
1 square foot (sq. ft.)	=	144 square inches (sq. in.)
1 square mile (sq. mi.)	=	640 acres
1 acre	=	43,560 square feet (sq. ft.)

ENGLISH VOLUME MEASURE FOR SOLIDS

1 cubic yard (cu. yd.)	=	27 cubic feet (cu. ft.)
1 cubic foot (cu. ft.)	=	1,728 cubic inches (cu. in.)

ENGLISH VOLUME MEASURE FOR FLUIDS

1 quart (qt.)	=	2 pints (pt.)
1 gallon (gal.)	=	4 quarts (qt.)

ENGLISH VOLUME MEASURE EQUIVALENTS

1 gallon (gal.)	=	0.133681 cubic foot (cu. ft.)
1 gallon (gal.)	=	231 cubic inches (cu. in.)

SI METRICS STYLE GUIDE

SI metrics is derived from the French name Le Systeme International d'Unites. The metric unit names are already in accepted practice. SI metrics attempts to standardize the names and usages so that students of metrics will have a universal knowledge of the application of terms, symbols, and units.

The English system of mathematics (used in the United States) has always had many units in its weights and measures tables that were not applied to everyday use. For example, the pole, perch, furlong, peck, and scruple are not used often. These measurements, however, are used to form other measurements and it has been necessary to include the measurements in the tables. Including these measurements aids in the understanding of the orderly sequence of measurements greater or smaller than the less frequently used units.

The metric system also has units that are not used in everyday application. Only by learning the lesser-used units is it possible to understand the order of the metric system. SI metrics, however, places an emphasis on the most frequently used units.

In using the metric system and writing its symbols, certain guidelines are followed. For the students' reference, some of the guidelines are listed.

1. In using the symbols for metric units, the first letter is capitalized only if it is derived from the name of a person.

SAMPLE:	UNIT	SYMBOL	UNIT	SYMBOL
	meter	m	Newton	N (named after Sir Isaac Newton)
	gram	g	degree Celsius	°C (named after Anders Celsius)
EXCEPTION:	The symbol for liter is L. This is used to distinguish it from the number one (1).			

2. Prefixes are written with lowercase letters.

SAMPLE:	PREFIX	UNIT	SYMBOL
	centi	meter	cm
	milli	gram	mg
EXCEPTIONS:	PREFIX	UNIT	SYMBOL
	tera	meter	Tm (used to distinguish it from the metric tonne,t)
	giga	meter	Gm (used to distinguish it from gram, g)
	mega	gram	Mg (used to distinguish it from milli, m)

3. Periods are not used in the symbols. Symbols for units are the same in the singular and the plural (no "s" is added to indicate a plural).

SAMPLE: 1 mm *not* 1 mm. 3 mm *not* 3 mms

4. When referring to a unit of measurement, symbols are not used. The symbol is used only when a number is associated with it.

SAMPLE: The length of the room is expressed in meters. *not* The length of the room is expressed in m. (*The length of the room is 25 m* is correct.)

5. When writing measurements that are less than one, a zero is written before the decimal point.

SAMPLE: 0.25 m *not* .25 m

6. Separate the digits in groups of three, counting from the decimal point to the left and to the right. A space is left between the groups of digits.

   SAMPLE: 5 179 232 mm *not* 5,179,232 mm 0.566 23 mg *not* 0.56623 mg
   1 346.098 7 L *not* 1,346.0987 L

   A space is also left between the digits and the unit of measure.

   SAMPLE: 5 179 232 mm *not* 5 179 232mm

7. Symbols for area measure and volume measure are written with exponents.

   SAMPLE: 3 cm^2 *not* 3 sq. cm 4 km^3 *not* 4 cu. km

8. Metric words with prefixes are accented on the first syllable. In particular, kilometer is pronounced "kill'-o-meter." This avoids confusion with words for measuring devices that are generally accented on the second syllable, such as thermometer (ther-mom'-e-ter).

METRIC RELATIONSHIPS

The base units in SI metrics include the meter and the gram. Other units of measure are related to these units. The relationship between the units is based on powers of ten and uses these prefixes:

kilo (1,000) hecto (100) deka (10) deci (0.1) centi (0.01) milli (0.001)

These tables show the most frequently used units with an asterisk (*).

METRIC LENGTH

10 millimeters (mm)*	=	1 centimeter (cm)*
10 centimeters (cm)	=	1 decimeter (dm)
10 decimeters (dm)	=	1 meter (m)*
10 meters (m)	=	1 dekameter (dam)
10 dekameters (dam)	=	1 hectometer (hm)
10 hectometers (hm)	=	1 kilometer (km)*

- To express a metric length unit as a smaller metric length unit, multiply by a positive power of ten such as 10, 100, 1,000, 10,000 etc.
- To express a metric length unit as a larger metric length unit, multiply by a negative power of ten such as 0.1, 0.001, 0.001, 0.000,1, etc.

METRIC AREA MEASURE

100 square millimeters (mm^2)	=	1 square centimeter (cm^2)*
100 square centimeters (cm^2)	=	1 sqaure decimeter (dm^2)
100 square decimeters (cm^2)	=	1 square meter (m^2)*
100 square meters (m^2)	=	1 square dekameter (dam^2)
100 square dekameters (dam^2)	=	1 square hectometer (hm^2)*
100 square hectometers (hm^2)	=	1 square kilometer (km^2)

- To express a metric area unit as a smaller metric area unit, multiply by 100, 10,000, 1,000,000, etc.
- To express a metric area unit as a larger metric area unit, multiply by 0.01, 0.000,1, 0.000,001, etc.

METRIC VOLUME MEASURE FOR SOLIDS

1 000 cubic millimeters (mm^3)	=	1 cubic centimeter (cm)*
1 000 cubic centimeters (cm^3)	=	1 cubic decimeter (dm^3)*
1 000 cubic decimeters (dm^3)	=	1 cubic meter (m^3)*
1 000 cubic meters (m^3)	=	1 cubic dekameter (dam^3)
1 000 cubic dekameters (dam^3)	=	1 cubic hectometer (hm^3)
1 000 cubic hectometers (hm^3)	=	1 cubic kilometer (km^3)

- To express a metric volume unit for solids as a smaller metric volume unit for solids, multiply by 1,000, 1,000,000, 1,000,000,000, etc.
- To express a metric volume unit for solids as a larger metric volume unit for solids, multiply by 0.001, 0.000,001, 0.000,000,001, etc.

METRIC VOLUME MEASURE FOR FLUIDS

10 milliliters (mL)*	=	1 centiliter (cL)
10 centiliters (cL)	=	1 deciliter (dL)
10 deciliters (dL)	=	1 liter (L)*
10 liters (L)	=	1 dekaliter (daL)
10 dekaliters (daL)	=	1 hectoliter (hL)
10 hectoliters (hL)	=	1 kiloliter (kL)

- To express a metric volume unit for fluids as a smaller metric volume unit for fluids, multiply by 10, 100, 1,000, 10,000, etc.
- To express a metric volume unit for fluids as a larger metric volume unit for fluids, multiply by 0.1, 0.01, 0.001, 0.000,1, etc.

METRIC VOLUME MEASURE EQUIVALENTS

1 cubic decimeter (dm^3)	=	1 liter (L)
1 000 cubic centimeters (cm^3)	=	1 liter (L)
1 cubic centimeter (cm^3)	=	1 milliliter (mL)

METRIC MASS MEASURE

10 milligrams (mg)*	=	1 centigram (cg)
10 centigrams (cg)	=	1 decigram (dg)
10 decigrams (dg)	=	1 gram (g)*
10 grams (g)	=	1 dekagram (dag)
10 dekagrams (dag)	=	1 hectogram (hg)
10 hectograms (hg)	=	1 kilogram (kg)*
1 000 kilograms (kg)	=	1 megagram (Mg)*

- To express a metric mass unit as a smaller metric mass unit, multiply by 10, 100, 1,000, 10,000, etc.
- To express a metric mass unit as a larger metric mass unit, multiply by 0.1, 0.01, 0.001, 0.000,1, etc.

Metric measurements are expressed in decimal parts of a whole number. For example, one-half millimeter is written as 0.5 mm.

In calculating with the metric system, all measurements are expressed using the same prefixes. If answers are needed in millimeters, all parts of the problem should be expressed in millimeters before the final solution is attempted. Diagrams that have dimensions in different prefixes must first be expressed using the same unit.

ENGLISH-METRIC EQUIVALENTS

LENGTH MEASURE

1 inch (in)	=	25.4 millimeters (mm)
1 inch (in)	=	2.54 centimeters (cm)
1 foot (ft)	=	0.304,8 meter (m)
1 yard (yd)	=	0.914,4 meter (m)
1 mile (mi)	≈	1.609 kilometers (km)
1 millimeter (mm)	≈	0.039,37 inch (in)
1 centimeter (cm)	≈	0.393,70 inch (in)
1 meter (m)	≈	3.280,84 feet (ft)
1 meter (m)	≈	1.093,61 yards (yd)
1 kilometer (km)	≈	0.621,37 mile (mi)

AREA MEASURE

1 square inch (sq in)	=	645.16 square millimeters (mm^2)
1 square inch (sq in)	=	6.451,6 square centimeters (cm^2)
1 square foot (sq ft)	≈	0.092,903 square meter (m^2)
1 square yard (sq yd)	≈	0.836,127 square meter (m^2)
1 square millimeter (mm^2)	≈	0.001,550 square inch (sq in)
1 square centimeter (cm^2)	≈	0.155,00 square inch (sq in)
1 square meter (m^2)	≈	10.763,910 square feet (sq ft)
1 square meter (m^2)	≈	1.195,99 square yards (sq yd)

VOLUME MEASURE FOR SOLIDS

1 cubic inch (cu in)	=	16.387,064 cubic centimeters (cm^3)
1 cubic foot (cu ft)	≈	0.028,317 cubic meter (m^3)
1 cubic yard (cu yd)	≈	0.764,555 cubic meter (m^3)
1 cubic centimeter (cm^3)	≈	0.061,024 cubic inch (cu in)
1 cubic meter (m^3)	≈	35.314,667 cubic feet (cu ft)
1 cubic meter (m^3)	≈	1.307,951 cubic yards (cu yd)

VOLUME MEASURE FOR FLUIDS

1 gallon (gal)	≈	3,785.411 cubic centimeters (cm^3)
1 gallon (gal)	≈	3.785,411 litres (L)
1 quart (qt)	≈	0.946,353 liter (L)
1 ounce (oz)	≈	29.573,530 cubic centimeters (cm^3)
1 cubic centimeter (cm^3)	≈	0.000,264 gallon (gal)
1 liter (L)	≈	0.264,172 gallon (gal)
1 litre (L)	≈	1.056,688 quarts (qt)
1 cubic centimeter (cm^3)	≈	0.033,814 ounce (oz)

MASS MEASURE

1 pound (lb)	≈	0.453,592 kilogram (kg)
1 pound (lb)	≈	453.592,37 grams (g)
1 ounce(oz)	≈	28.349,523 grams (g)
1 ounce(oz)	≈	0.028,350 kilogram (kg)
1 kilogram (kg)	≈	2.204,623 pounds (lb)
1 gram (g)	≈	0.002,205 pound (lb)
1 kilogram (kg)	≈	35.273,962 ounces (oz)
1 gram (g)	≈	0.035,274 ounce (oz)

WEIGHTS OF COMMON METALS

Aluminum	=	0.0975 lb/in^3
Brass	=	0.3105 lb/in^3
Copper	=	0.321 lb/in^3
Steel	=	0.283 lb/in^3

DECIMAL EQUIVALENTS

Fraction	Decimal Equivalent English (in.)	Decimal Equivalent Metric (mm)	Fraction	Decimal Equivalent English (in.)	Decimal Equivalent Metric (mm)
1/64	0.015625	0.3969	33/64	0.515625	13.0969
1/32	0.03125	0.7938	17/32	0.53125	13.4938
3/64	0.046875	1.1906	35/64	0.546875	13.8906
1/16	0.0625	1.5875	9/16	0.5625	14.2875
5/64	0.078125	1.9844	37.64	0.578125	14.6844
3/32	0.09375	2.3813	19/32	0.59375	15.0813
7/64	0.109375	2.7781	39/64	0.609375	15.4781
1/8	0.1250	3.1750	5/8	0.6250	15.8750
9/64	0.140625	3.5719	41/64	0.640625	16.2719
5/32	0.15625	3.9688	21/32	0.65625	16.6688
11/64	0.171875	4.3656	43/64	0.671875	17.0656
3/16	0.1875	4.7625	11/16	0.6875	17.4625
13/64	0.203125	5.1594	45/64	0.703125	17.8594
7/32	0.21875	5.5563	23/32	0.71875	18.2563
15/64	0.234375	5.9531	47/64	0.734375	18.6531
1/4	0.250	6.3500	3/4	0.750	19.0500
17/64	0.265625	6.7469	49/64	0.765625	19.4469
9/32	0.28125	7.1438	25/32	0.78125	19.8438
19/64	0.296875	7.5406	51/64	0.796875	20.2406
5/16	0.3125	7.9375	13/16	0.8125	20.6375
21/64	0.328125	8.3384	53/64	0.828125	21.0344
11/32	0.34375	8.7313	27/32	0.84375	21.4313
23/64	0.359375	9.1281	55/64	0.859375	21.8281
3/8	0.3750	9.5250	7/8	0.8750	22.2250
25/64	0.390625	9.9219	57/64	0.890625	22.6219
13/32	0.40625	10.3188	29/32	0.90625	23.0188
27/64	0.421875	10.7156	59/64	0.921875	23.4156
7/16	0.4375	11.1125	15/16	0.9375	23.8125
29/64	0.453125	11.5094	61/64	0.953125	24.2094
15/32	0.46875	11.9063	31/32	0.96875	24.6063
31/64	0.484375	12.3031	63/64	0.984375	25.0031
1/2	0.500	12.7000	1	1.000000	25.4000

SECTION III
REFERENCE MATERIAL FOR MACHINISTS

TAP, DIE, AND THREADING TOOL DESIGNATIONS

This form is used to designate taps, dies, and threading tools.

NOMINAL SIZE — THREADS PER INCH — THREAD SERIES

THREAD SERIES (MOST COMMON)

SYMBOL	THREAD DESIGNATION
NC	American National Coarse Unified Coarse*
NF	American National Fine Unified Fine*
NEF	American National Extra Fine Unified Extra Fine*
NS	American National Thread - Special Unified Thread - Special*
	*Taps are not marked with "U" but with the symbol for the corresponding American Standard thread form.

SAMPLES: $\frac{1''}{4}$ - 20 NC

$\frac{5''}{8}$ - 18 NF

In addition, this form is used to designate ground thread taps.

NOMINAL SIZE — THREADS PER INCH THREAD SERIES — [GROUND TAP DESIGNATION: G SERIES] [STANDARD TOLERANCE RANGE: PITCH DIAMETER LIMITS]

SERIES

SYMBOL	DESIGNATION
H	high taps, tolerance above basic
L	low taps, tolerance below basic

PITCH DIAMETER LIMITS

SYMBOL	AMOUNT ABOVE OR BELOW BASIC SIZE
1	0.0000 in. – 0.0005 in.
2	0.0005 in. – 0.0010 in.
3	0.0010 in. – 0.0015 in.
4	0.0015 in. – 0.0020 in.
5	0.0020 in. – 0.0025 in.
6	0.0025 in. – 0.0030 in.

STANDARD TOLERANCE RANGES (MOST COMMON) FOR TAPS TO ONE INCH IN DIAMETER, INCLUSIVE

SYMBOL	DESIGNATION
L_1	basic to basic minus 0.0005 in.
H_1	basic to basic plus 0.0005 in.
H_2	basic plus 0.0005 in. to basic plus 0.0010 in.
H_3	basic plus 0.0010 in. to basic plus 0.0015 in.
H_4	basic plus 0.0015 in. to basic plus 0.0020 in.
H_5	basic plus 0.0020 in. to basic plus 0.0025 in.
H_6	basic plus 0.0025 in. to basic plus 0.0030 in.

SAMPLES: $\frac{5''}{8}$ - 11 NC - GH_2 (pitch diameter limits are 0.0005 in. to 0.0010 in. above basic)

$\frac{1''}{4}$ - 20 NF - GL_1 (pitch diameter limits are 0.0000 in. to 0.0005 in. below basic)

THREAD DESIGNATIONS

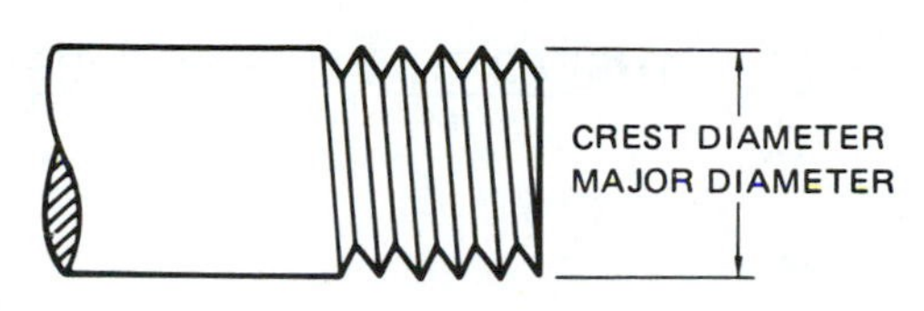

EXTERNAL THREAD (BOLT)

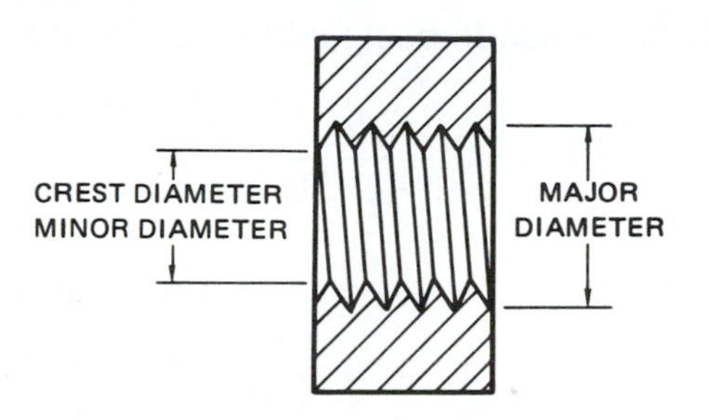

INTERNAL THREAD (NUT)

BRITISH STANDARD

This form is used to designate screw threads belonging to the British Standard series.

METRIC THREAD DESIGNATION			TOLERANCE CLASS (PITCH DIAMETER)		TOLERANCE CLASS (CREST DIAMETER)	
M	DIAMETER	PITCH	GRADE	POSITION (ALLOWANCE)	GRADE	POSITION (ALLOWANCE)

TOLERANCE GRADE

TYPE OF FIT	EXTERNAL THREADS		INTERNAL THREADS	
	PITCH DIAMETER	MAJOR (CREST) DIAMETER	PITCH DIAMETER	MINOR (CREST) DIAMETER
very fine	3	------	------	------
fine	4	4	4	4
fine	5	------	5	5
medium	6	6	6	6
medium	7	------	7	7
coarse	8	8	8	8
free fit	9	------	------	------

TOLERANCE POSITION (ALLOWANCE)

TYPE OF ALLOWANCE	EXTERNAL THREADS	INTERNAL THREADS
no allowance	h	H
small allowance (average)	g	G
large allowance	e	------

SAMPLES: M12 x 1.75 - 5H6H (internal thread)

M8 x 1.25 - 6 g (external thread)

Note: If the pitch and crest diameter tolerances are the same, the symbol is used only once.

AMERICAN STANDARD

This form is used to designate screw threads belonging to the American Standard series.

MAJOR DIAMETER – THREADS PER INCH THREAD SERIES – THREAD FIT TYPE OF THREAD

THREAD SERIES (MOST COMMON)

SYMBOL	THREAD DESIGNATION
UN	Unified Constant Pitch Series
UNC	Unified Coarse
UNF	Unified Fine
UNEF	Unified Extra Fine
UNS	Unified Thread – Special

THREAD FIT
(EXTERNAL AND INTERNAL THREADS)

TYPE OF FIT	SYMBOL
loose	1
medium	2
tight	3

TYPE OF THREAD

THREAD	SYMBOL
external	A
internal	B

SAMPLES: $\frac{3''}{8}$ - 16 UNC - 2A (external thread)

$\frac{7}{16}$ - 20 UNF - 1B (internal thread)

THREAD CALCULATIONS

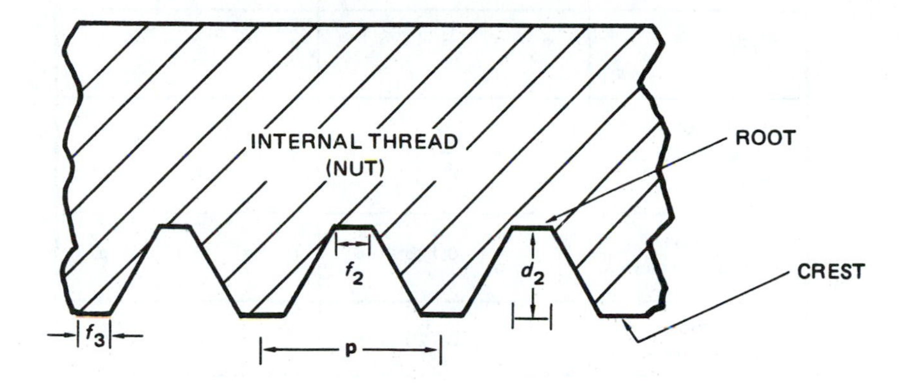

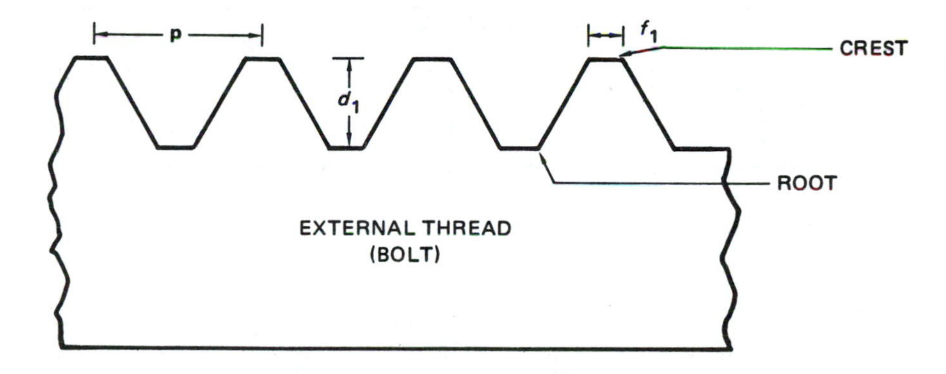

SYMBOL	MEANING
p	pitch
n	number of threads per inch or millimetre
f_1	flat at crest, external crest
f_2	flat at root, internal thread
f_3	flat at crest, internal crest
d_1	depth, external thread
d_2	depth, internal thread

TO FIND	UNIFIED	METRIC
Pitch	$p = \frac{1}{n}$	$p = \frac{1}{n}$
Number of threads per inch or millimetre	$n = \frac{1}{p}$	$n = \frac{1}{p}$
Flat at crest, external thread	$f_1 = 0.125 \times p$	$f_1 = 0.125 \times p$
Flat at root, internal thread	$f_2 = 0.125 \times p$	$f_2 = 0.125 \times p$
Flat at crest, internal thread	$f_3 = 0.250 \times p$	$f_3 = 0.250 \times p$
Depth, external thread	$d_1 = 0.61343 \times p$	$d_1 = 0.613\ 43 \times p$
Depth, internal thread	$d_2 = 0.54127 \times p$	$d_2 = 0.541\ 27 \times p$

SYMBOL	MEANING
M	measurement over wires
D	major (outside) diameter
p	pitch
w	wire size

THREE WIRE METHOD

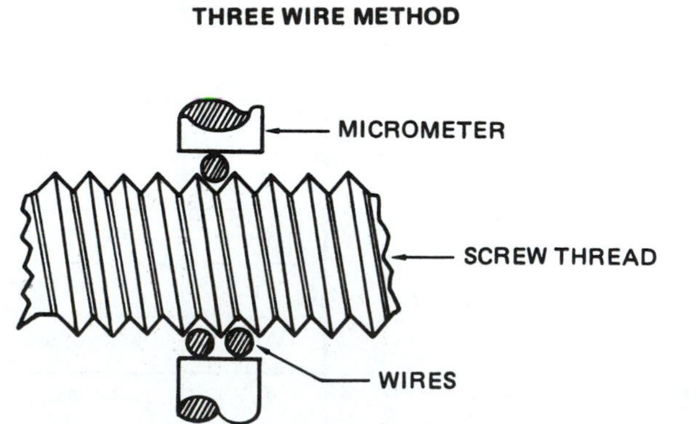

$$M = [D - (1.5155 \times p)] + 3w$$

$$w = 0.57735 \times p \text{ (best)}$$

$$w = 0.5600 \times p \text{ (minimum)}$$

$$w = 0.9000 \times p \text{ (maximum)}$$

TAP DRILL SIZES - COMMON SCREW AND BOLT SIZES

SIZE/TPI	OD	TAP DRILL	DRILL OD
0 x 80 UNF	.060	3/64	.0469
2 x 56 UNC	.086	#50	.0700
4 x 40 UNC	.112	#43	.0890
6 x 32 UNC	.138	#36	.1065
8 x 32 UNC	.164	#29	.1360
10 x 24 UNC	.190	#25	.1495
1/4 x 20 UNC	.250	#7	.2010
1/4 x 28 UNF	.250	#3	.2130
5/16 x 18 UNC	.313	F	.2570
5/16 x 24 UNF	.313	L	.2720
3/8 x 16 UNC	.375	5/16	.3125
3/8 x 24 UNF	.375	Q	.3320
1/2 x 13 UNC	.500	27/64	.4129
1/2 x 20 UNF	.500	29/64	.4531
5/8 x 11 UNC	.625	17/32	.5312
5/8 x 18 UNF	.625	37/64	.5781
3/4 x 10 UNC	.750	21/32	.6562
3/4 x 16 UNF	.750	11/16	.6875
1 x 8 UNC	1.00	7/8	.8750
1 x 14 UNF	1.00	15/16	.9375
Tap Drill sizes based on 75% thread			

GEAR CALCULATIONS

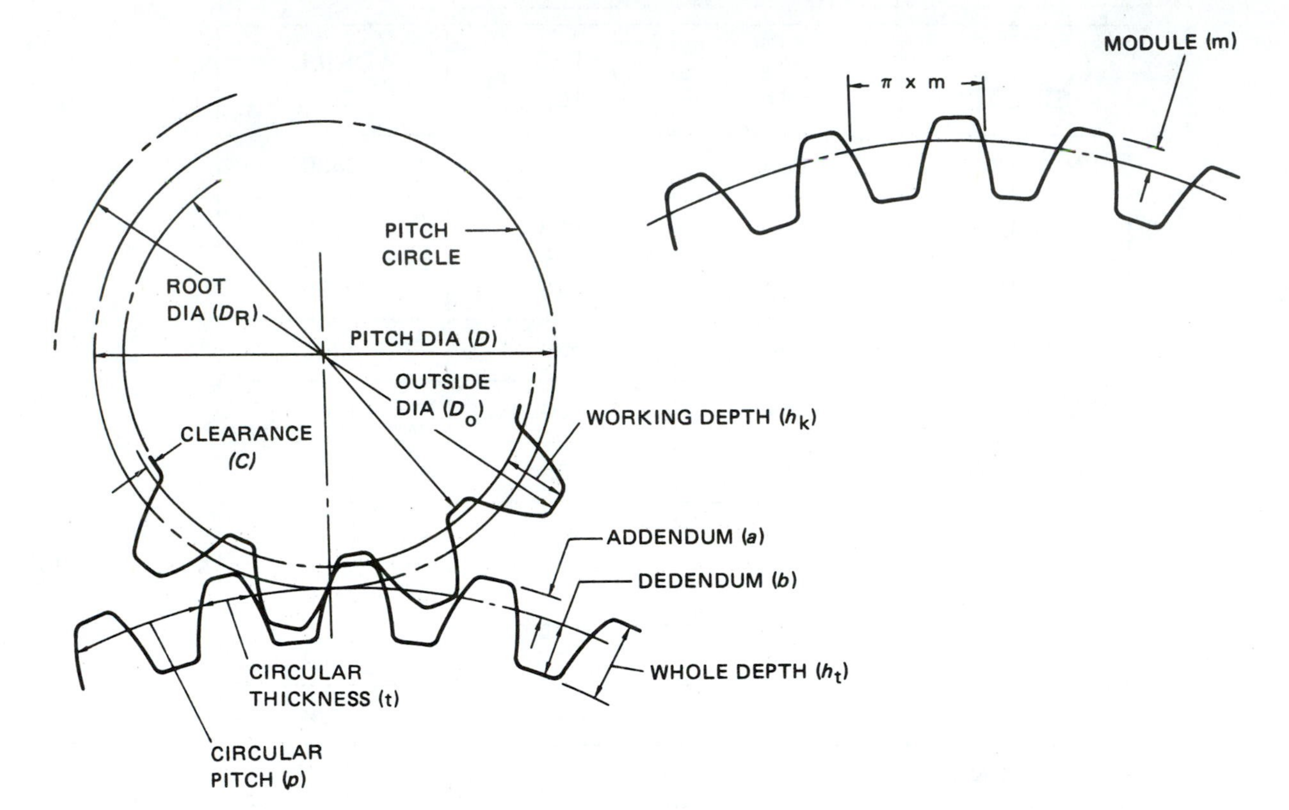

SYMBOL	MEANING
D	pitch diameter
D_o	outside diameter
D_R	root diameter
P	diametral pitch
p	circular pitch
t	circular thickness

SYMBOL	MEANING
a	addendum
b	dedendum
c	clearance
h_t	whole depth
h_k	working depth
n	number of teeth
m	module

TO FIND	AMERICAN NATIONAL STANDARD	METRIC
Module	- - - - - - - - - -	$m = \frac{D}{n}$
Diametral pitch	$P = \frac{n}{D}$	$p = \frac{1}{m}$
Pitch diameter	$D = \frac{n}{P}$ $D = \frac{D_o \times n}{(n + 2)}$	$D = n \times m$
Number of teeth (expressed as a whole number	$n = P \times D$	$n = \frac{D}{m}$
Addendum	$a = \frac{1}{P}$	$a = m$
Dedendum	$b = \frac{1.250}{P}$	$b = 1.250 \times m$
Clearance (preferred)	$c = \frac{0.250}{P}$	$c = 0.250 \times m$
Clearance (minimum)	$c = \frac{0.157}{P}$	$c = 0.157 \times m$
Circular thickness-Basic	$t = \frac{1.5708}{P}$	$t = 1.570\,8 \times m$
Root diameter	$D_R = \frac{(n - 2.5)}{P}$ $D_R = D - (2 \times b)$	$D_R = D - (2.5 \times m)$
Outside diameter	$D_o = \frac{(n + 2)}{P}$ $D_o = D + (2 \times a)$	$D_o = m \times (n + 2)$ $D_o = D + (2 \times m)$
Whole depth (preferred)	$h_t = \frac{2.250}{P}$ $h_t = a + b$	$h_t = a + b$
Working depth	$h_k = \frac{2}{P}$ $h_k = a + b - c$	$h_k = 2 \times a$
Circular pitch	$p = \frac{3.1416}{P}$	$p = 3.141\,6 \times m$

TAPERS

SYMBOL	MEANING
tpi	taper per inch
tpf	taper per foot
D	diameter, larger end
d	diameter, small end
L	length (in inches)

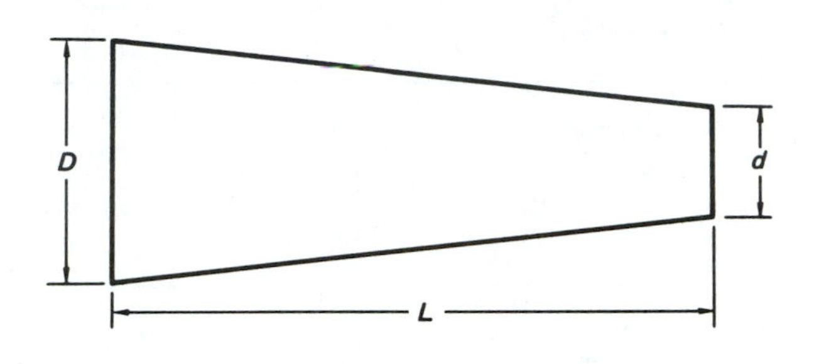

TO FIND	KNOWN	FORMULA
tpf	*tpi*	$tpf = tpi \times 12$
	D, d, L	$tpf = 12\left(\frac{D - d}{L}\right)$
tpi	*tpf*	$tpi = \frac{tpf}{12}$
	D, d, L	$tpi = \frac{D - d}{L}$
d	*D, L, tpf*	$d = D - \left[L\left(\frac{tpf}{12}\right)\right]$
D	*d, L, tpf*	$D = d + \left[L\left(\frac{tpf}{12}\right)\right]$
L	*D, d, tpf*	$L = 12\left(\frac{D - d}{tpf}\right)$

BEND ALLOWANCES

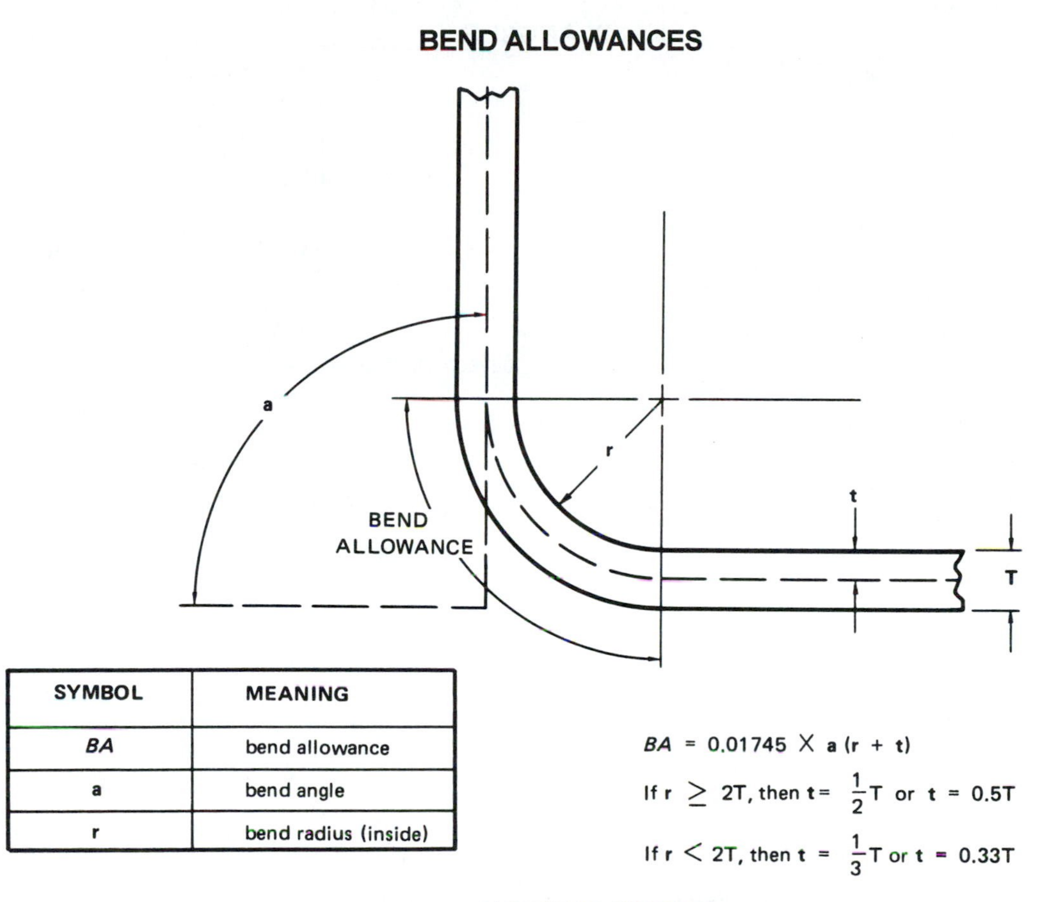

SYMBOL	MEANING
BA	bend allowance
a	bend angle
r	bend radius (inside)

$BA = 0.01745 \times a\,(r + t)$

If $r \geq 2T$, then $t = \frac{1}{2}T$ or $t = 0.5T$

If $r < 2T$, then $t = \frac{1}{3}T$ or $t = 0.33T$

CUTTING SPEEDS

SYMBOL	MEANING	AMERICAN STANDARD UNITS	METRIC UNITS
V	cutting speed	feet per minute (fpm)	metres per minute (m/min)
D	diameter	inches (in.)	millimetres (mm)
N	spindle speed	revolutions per minute (rpm)	revolutions per minute (r/min)

TO FIND	AMERICAN STANDARD UNITS	METRIC UNITS
N	$N = \frac{12 \times V}{3.1416 \times D}$	$N = \frac{1\,000 \times V}{3.141\,6 \times D}$
V	$V = \frac{3.1416 \times D \times N}{12}$	$V = \frac{3.141\,6 \times D \times N}{1\,000}$
D	$D = \frac{V \times 12}{3.1416 \times N}$	$D = \frac{V \times 1\,000}{3.141\,6 \times N}$

HELIX ANGLES

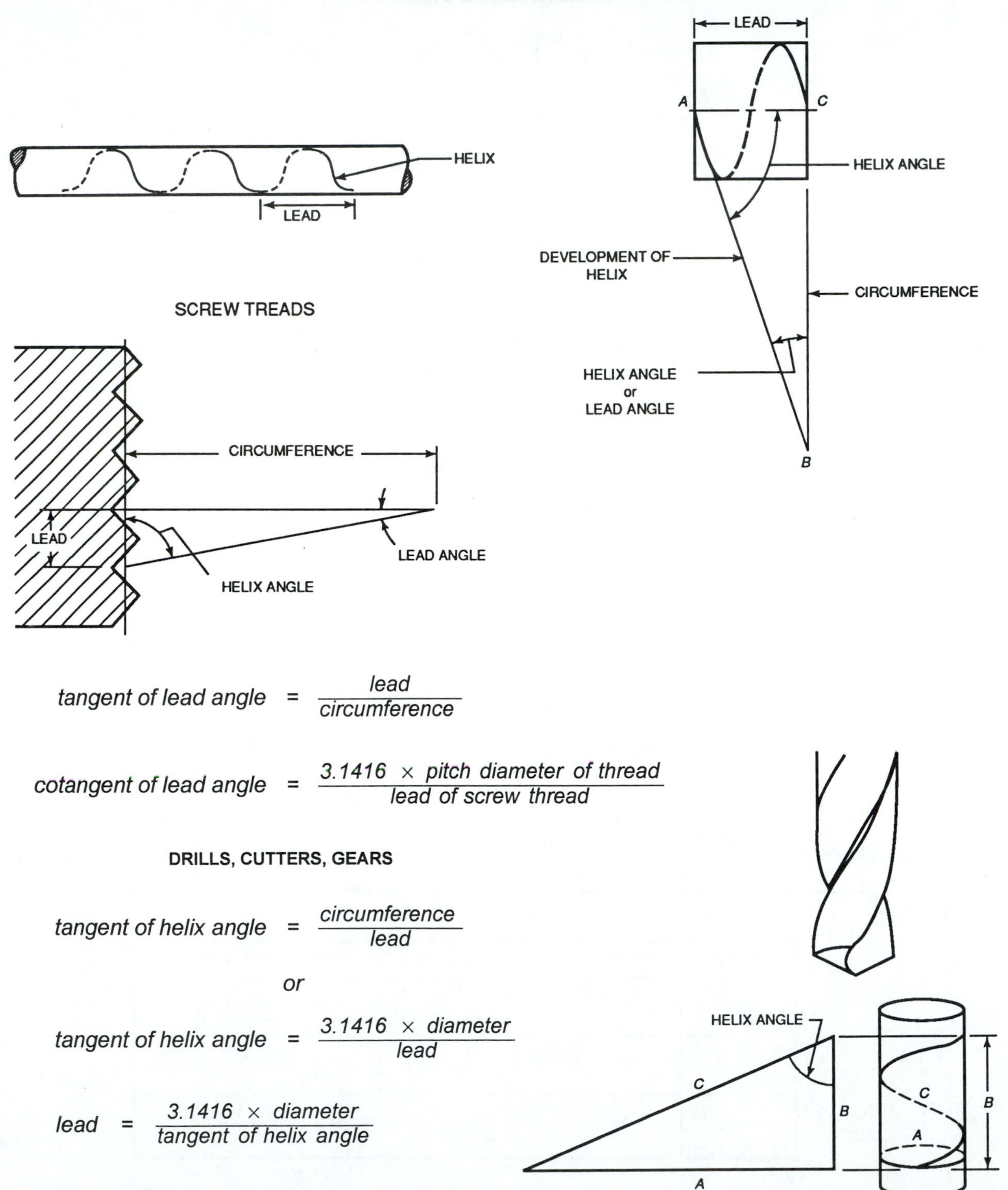

$$\text{tangent of lead angle} = \frac{\text{lead}}{\text{circumference}}$$

$$\text{cotangent of lead angle} = \frac{3.1416 \times \text{pitch diameter of thread}}{\text{lead of screw thread}}$$

DRILLS, CUTTERS, GEARS

$$\text{tangent of helix angle} = \frac{\text{circumference}}{\text{lead}}$$

or

$$\text{tangent of helix angle} = \frac{3.1416 \times \text{diameter}}{\text{lead}}$$

$$\text{lead} = \frac{3.1416 \times \text{diameter}}{\text{tangent of helix angle}}$$

MEASURING ANGLES WITH DISCS

DOVETAILS

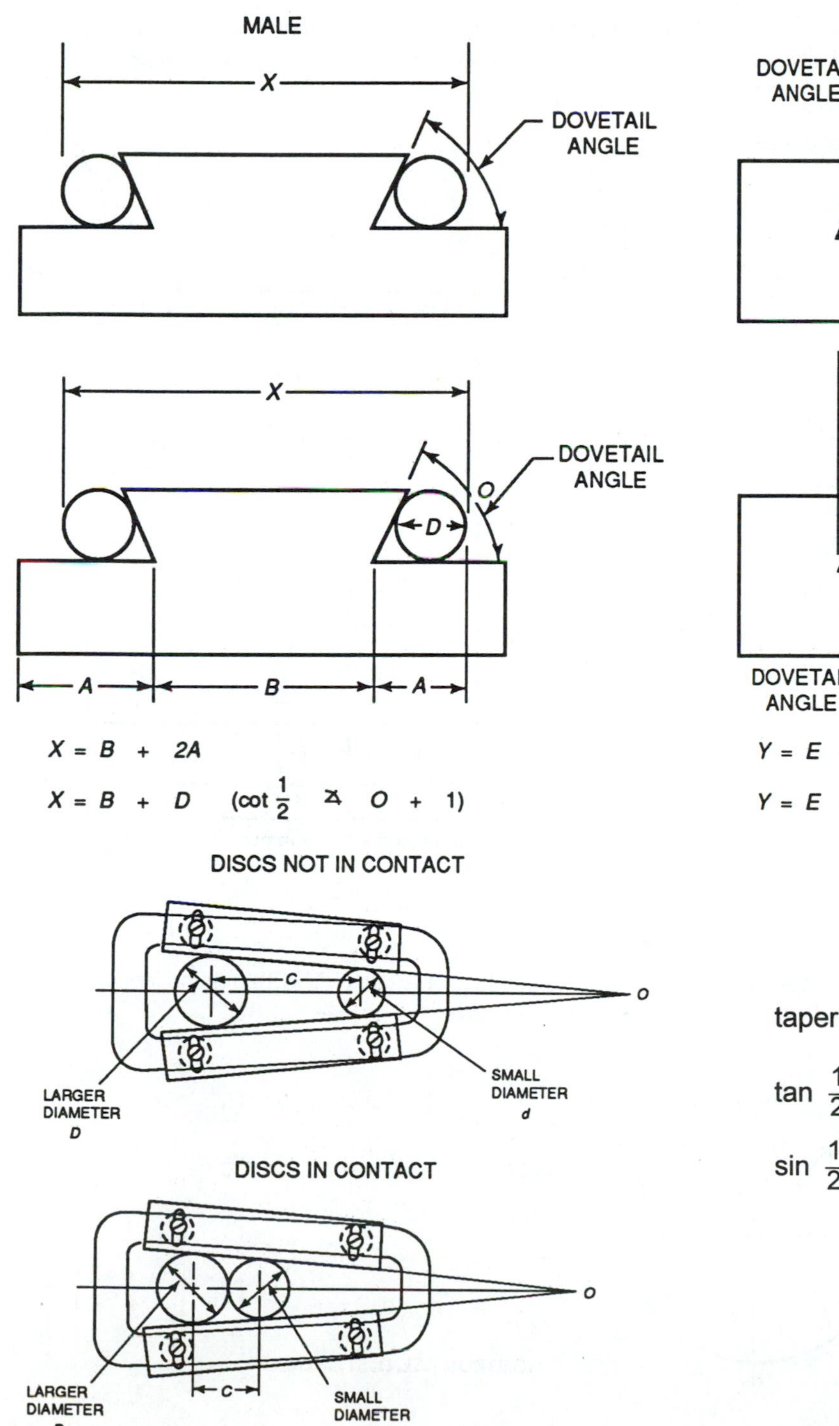

$X = B + 2A$

$X = B + D\ (\cot \frac{1}{2} \angle O + 1)$

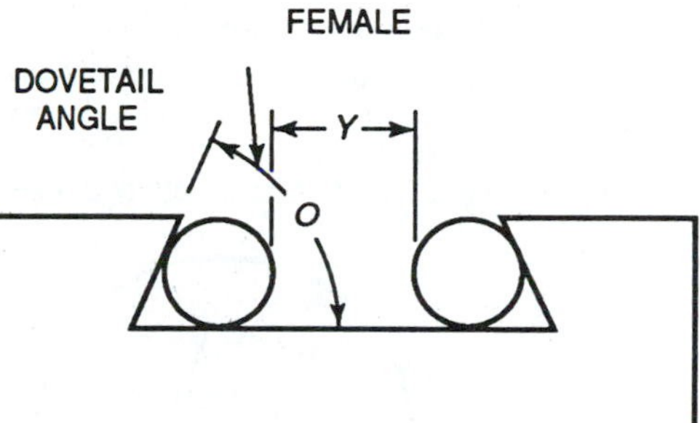

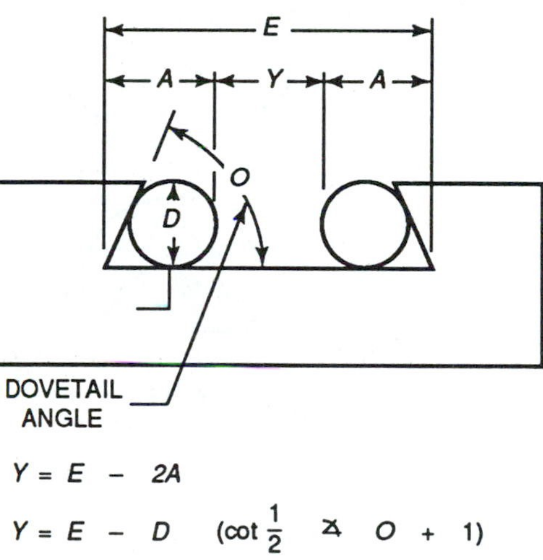

$Y = E - 2A$

$Y = E - D\ (\cot \frac{1}{2} \angle O + 1)$

$$\text{taper per inch (millimeter)} = \frac{D - d}{C}$$

$$\tan \frac{1}{2} \angle O = \frac{\text{taper per foot}}{24}$$

$$\sin \frac{1}{2} \angle O = \frac{D - d}{2C}$$

INDEXING HOLES

USING CHORDS

This table lists the chord constants for a diameter of one millimeter/one inch. To use the chart, select the constant for the desired number of divisions and multiply it by the diameter of the bolt circle. This results in the length of the chord to be used.

A *chord* is a straight line that connects two points of an arc.

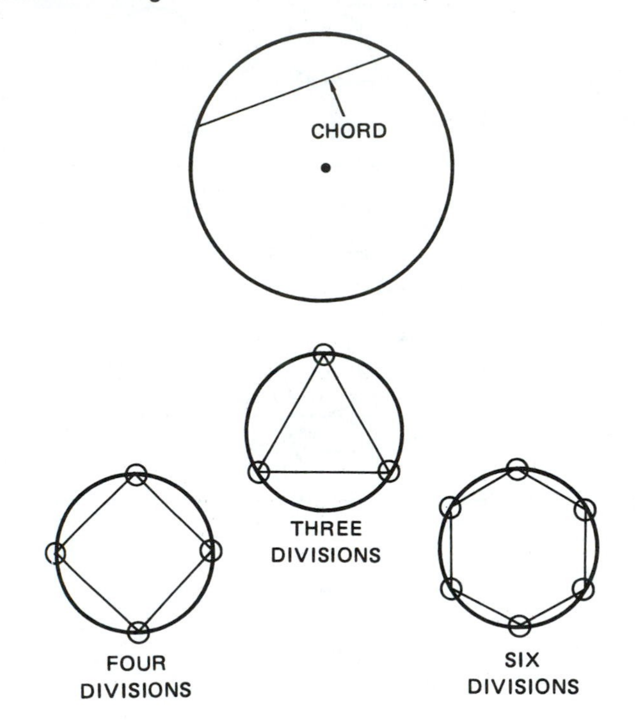

DIVISIONS	DEGREE	CONSTANT*
3	120°	0.8660
4	90°	0.7071
5	72°	0.5878
6	60°	0.5000
8	45°	0.3827
10	36°	0.3090
12	30°	0.2588
15	24°	0.2079
18	20°	0.1736
20	18°	0.1564
24	15°	0.1305
30	12°	0.1045
32	$11\frac{1}{4}$°	0.0980
36	10°	0.0872

*Based on a hole diameter of 1 mm/1 in.

USING HORIZONTAL AND VERTICAL DISTANCES

Measuring horizontal and vertical distances can be used to locate holes on a circle. The beginning point is the center of the circle. The distance is constants times the diameter of the circle.

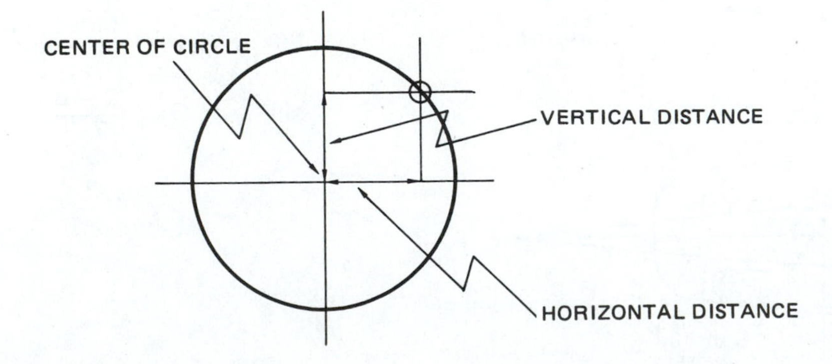

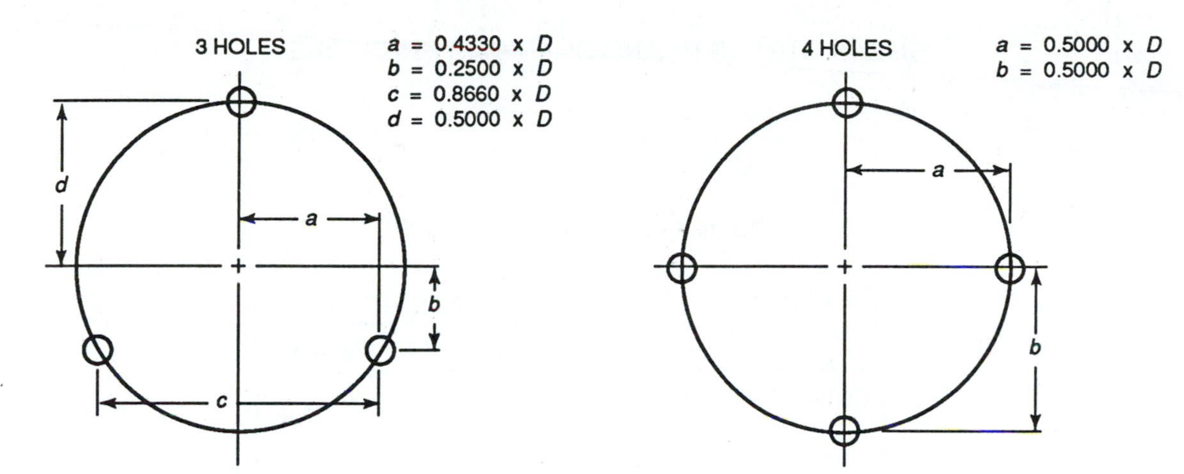

Fig. A-17

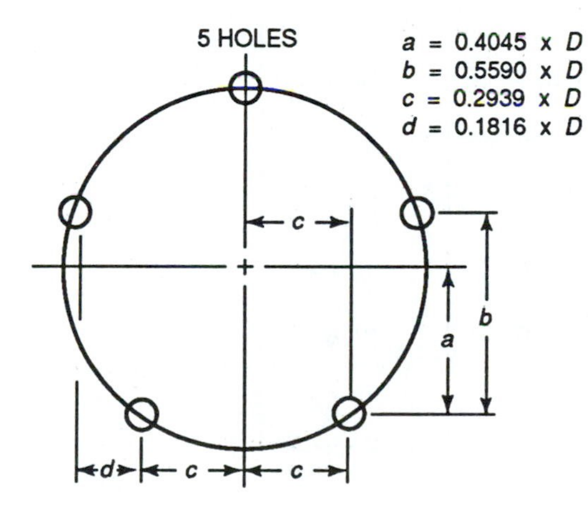

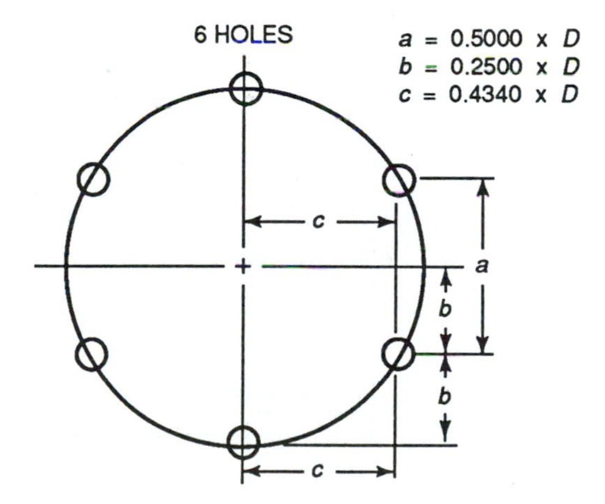

Fig. A-18

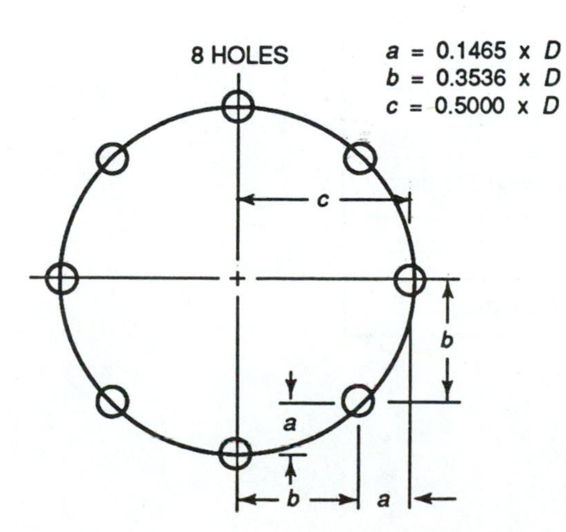

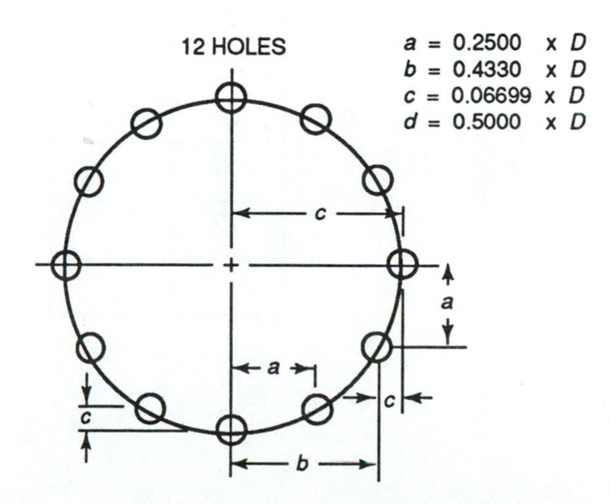

SHEAR STRENGTH AND BLANKING PRESSURE

APPROXIMATE SHEAR STRENGTHS

MATERIAL	SHEAR STRENGTH	
	lb./sq. in.	kg/mm^2
Low Carbon Steel *1020*	43,000	30.23
Med Carbon Steel *1050*	81,000	56.94
High Carbon Steel *1095*	105,000	73.82
Alloy Steels		
2340	123,000	86.47
3250	140,000	98.42
4130	80,000	56.24
Aluminum Alloys	28,000	19.68
Brass	35,000	24.61
Magnesium Alloys	25,000	17.58

SYMBOL	MEANING
BP	blanking pressure
P	perimeter of the part
t	thickness of the part
s	shear strength

$$BP = P \times t \times s$$

GLOSSARY

Adapter plate - A steel plate fastened to a machine that enables the machine to hold fixtures, workpieces, or dies, for production machining or machining beyond the capacity of the machine.

Addendum - The radial distance between the pitch circle and the top of the tooth.

Alloy - A mixture of two or more metals combined together to form a new metal.

Angle cutter - A milling cutter for a horizontal or vertical miller used to mill an angle on a workpiece without requiring the spindle axis to be changed.

Angle plate - A right-angle accurately machined plate used to hold workpieces, with clamps, for machining purposes; a sine bar designed and used to machine compound angles.

Arbor - A shaft on which a revolving cutting tool is mounted; a spindle on a cutting machine that holds the work to be cut.

Automatic screw machine - A machine used to accurately produce large quantities of parts at a high rate of speed and without the use of a full-time operator.

Babbitt - A metal used as an antifriction lining for bearings. It is composed of antimony, tin, and copper.

Bar - The form in which metal is received in a shop. It can be obtained in any shape, such as square or round.

Bar stock - Bars or other material that are to be worked on.

Bearing - A part in which a shaft revolves.

Bench plate - A cast iron plate that has been accurately machined to a smooth surface and is used for layout work.

Bend allowance - The amount of stock required to make a bend through a specific angle and over a specific radius.

Blanking die - A press-working tool that is used to stamp metal parts. A strip or sheet of metal is fed into a machine that punches out one or more blanks at one time.

Bolt circle - A circular centerline locating the centers of holes that are positioned around a common center point.

Bore - To make or enlarge a hole with a single-point cutting tool.

Carburize - A process in which carbon is added to the surface of steel by heating the metal. Carburizing is frequently followed by quenching to produce a hardening case.

Case-hardening steel - A low-carbon steel that is hardened by carburizing the outer surface and then heat-treating the carburized parts.

Casting - A process by which a part is formed by pouring molten metal into a mold or container where it solidifies.

Center punch - A small hard-steel bar with a conical or blunt point that is used to mark the center of a hole that is to be drilled.

Chip - Material that is being removed from the workpiece by the machining process.

Chisel - A flat steel cutting tool that is beveled and sharpened on one end. It has many uses including the cutting of sheet metals, bolts, or keyways.

Chuck - A device used to secure a tool or piece of material into position on a lathe or other rotating machine.

Chucking - The process of placing a workpiece in a chuck.

Circular pitch - The length of the arc on a pitch circle measured between corresponding points on adjacent gear teeth.

Clearance - The distance by which one part clears another part. On gears, it is the amount by which the dedendum of one gear exceeds the addendum of the mating gear.

Clevis - A *U*-shaped connector used to connect a rod to a plate or a lever.

Cold-rolled steel - Steel that has been rolled while cold to produce smooth accurate stock that contains a specific amount of carbon.

Collar - A round flange or ring formed on a shaft by forging or locking to the shaft by means of a setscrew or sweating.

Cone pulley - A step pulley system used to provide various speed ratios between the motor and the headstock spindle.

Coolant - A fluid used in machining operations to cool work and remove chips.

Cored hole - A hole made by leaving a core in a piece of metal when molding. The core is removed and a neat hole results.

Cotter pin - a half-round metal strip bent into a pin that has ends that can be flared after insertion through a slot or a hole.

Countershaft - The smaller shaft and pulley located over a machine and driven by the main line shaft; used to change the direction of rotation.

Cover plate - A removable plate used to protect the operator from moving parts of the machine, to prevent oil from leaking from the machine. When removed it allows access into the machine to make machine repairs.

Crankpin - A cylindrical piece that forms the handle of a crank or to which the connecting rod is attached.

Crest - The top surface joining the two sides or flanks of a thread.

Dedendum - The radial distance between the pitch circle and the bottom of the tooth space.

Diametral pitch - The ratio of the number of teeth to the measure of the pitch diameter; number of gear teeth in each unit of pitch diameter.

Die - A device used for forming, impressing, stamping, bending, or cutting metals, or for cutting external threads.

Dovetail - A type of joint in which the pieces join in an interlocking fashion.

Dowel pin - A steel pin used to retain parts in a fixed position or to preserve alignment.

Driven gear or pulley - The gear or pulley that receives the power.

Driver gear or pulley - the gear or pulley that transmits the power.

End milling cutters - A milling cutter that is shank mounted and designed to cut on both its sides and end.

External thread - The thread on the outside of a part, such as on a screw or bolt.

Facing - A lathe operation where the ends of the work are made flat and smooth.

Flute milling - The removal of material by the use of the side of the milling cutter.

Gauge - An instrument with a graduated scale or dial for measuring or indicating quantity. Also used to indicate, by number, the thickness of sheet metal.

Gauge blocks - Rectangular pieces of tool steel that are hardened, ground, lapped, and finished accurately to tolerance within a few millionths of an inch. Gauge blocks are used as a practical reference standard for measuring length.

Gear - A toothed wheel used to transmit power or motion from one shaft to another shaft.

Gib - A metal shim, usually movable, that is installed to allow moving machine parts to be adjusted for wear.

Hacksaw - A metal U-shaped frame with a handle attached to one end that contains saw blades designed to cut metal. The blades are fastened to the frame with a screw and wing nut.

Helix - A curve formed on a cylinder by a point rotating around the surface at a constant angle.

Helix angle - The angle to which the table must be set or to which the cutter clearance must be ground or set to produce a given helical curve.

High speed steel - A type of tool steel that is made to withstand considerable heat and still retain its hardness. This type of steel is used for milling cutters, drills, and lathe tool bits.

Index - The process of dividing a cylindrical piece of work into equal parts in order to cut gear teeth or to drill equally spaced holes.

Index plate - A disc consisting of a series of equally spaced holes that is used to divide a portion of an index turn into accurate increments.

Internal thread - The thread on the inside of a part, such as on a nut.

Jig - A device that holds the work and guides the tool during a cutting operation. This device is used for the production of large quantities of identical parts.

Keyway - A groove cut in a shaft or in a hole of a gear to allow a key to be inserted.

Lead - The distance a screw moves axially in one revolution; the distance that a point moving around a cylindrical surface advances axially in one complete turn.

Lead angle - The measure of the inclination of a screw thread from a plane that is perpendicular to the screw thread axis.

Major diameter - The largest diameter of a straight screw thread.

Micrometer caliper - A length-measuring instrument that measures to within one thousandth inch or one hundredth millimeter.

Minor diameter - The smallest diameter of a straight screw thread.

Module - The ratio of the pitch diameter to the number of teeth; the size of the gear tooth.

Naval brass - A special type of metal that is stronger, tougher, and more corrosion resistant than commercial brass rods. It is used for studs and nuts, bushings, and screw machine parts.

Numerical control - A positioning-control system for machine tools and other kinds of industrial-production machines. Numerical values corresponding to desired positions of a tool are recorded on punched tape and other devices.

Offset of tailstock - The amount by which the tailstock is shifted off center to cut a taper on a lathe.

Parting tool - A thin offset-cutting tool that is used to cut material at specified lengths on a lathe.

Pinion gear - The smaller of a pair of gears; a gear with a small number of teeth designed to mesh with a larger gear.

Pitch - The distance between similar, equally spaced tooth surfaces measured in a given direction and along a given curve or line.

Pitch circle - A circle that has a radius that is equal to the distance from the gear axis to the pitch point.

Pitch diameter - The diameter of the pitch circle.

Profile gauge - A flat metal plate that is accurately machined and is used to determine the proper shape of a machined part along its axis.

Radial drill - A drill press with a spindle that is mounted in a radial arm supported on a column. This machine is especially useful when drilling several holes in larger and heavier pieces.

Reamer - A tool used for enlarging and/or finishing a hole to an accurate size and to a fine micro finish.

Rivet - A headed metal pin used to fasten materials together.

Root - The bottom surface joining adjacent sides or flanks of a thread or gear.

Screw thread - A raised portion produced on the outside of a cylinder or on the inside walls of a cylindrical hole after a thread-cutting operation has been completed.

Screw thread micrometer - A special type of micrometer used to measure the pitch diameter of a screw.

Shear angle - The angle on the cutting edge of a tool designed to cut materials with the least amount of force required. It can be found on tin snips, sheet metal shears, piercing and blanking dies.

Side milling cutters - A circular milling cutter that is arbor-mounted and designed to cut on both its circumference and sides.

Sine bar - A highly finished and accurate chrome steel part that rests on two cylindrical rollers whose centers are either five or ten inches apart. This instrument is primarily used to measure angles.

Slab cutter - A milling cutter with a large end diameter and carbide or high-speed steel inserts. It is used to remove large amounts of metal in a single cut.

Slitting saw - A thin milling saw that is primarily used for cut-off operations.

Splined shaft - A rotating shaft with a series of keyways milled into it. The keyways fit into a matching bushing that allows the shaft to be moved in or out, or up or down such as on the automatic feed on a miller or on a spindle of a drill press.

Step pulley drive - A pulley with a series of different diameters that is used with a flat or V-belt to vary the spindle speed of machines such as lathes and vertical millers. The pulleys are mounted in inverted sets with the large diameter matching up with the small diameter of the next pulley.

Stock - Commonly used and commercially available patterns and sizes of material, especially metal.

Stroke - The length of travel of the ram or piston of a machine.

Tailstock - The sliding head of a lathe that gives an outer bearing and support for work being turned on centers.

Taper - A uniform change in size.

Taper per foot - The difference in diameters, for any length, measured along the axis of the work and expressed in inches per foot.

Thrust washer - A type of washer that holds a rotating part from sideward movement in its bearing.

Tool crib - A room used to store tools, machine parts, and equipment. An attendant is usually in charge of the maintenance and inventory of the crib room.

Tool steel - Steel that is used primarily for making tools used in manufacturing and trades. It has specific properties and must withstand high specific loads.

Turret Lathe - A lathe that is provided with a revolvable or pivoted hold to carry several cutting tools in place of the tailstock. It is used for high-speed, high-volume production.

Twist drill - A type of drill with a helical fluted body that is used to drill holes.

Vernier calipers - A steel scale or bar that is used for accurate inside and outside measurements.

Vertical milling machine - A milling machine that has a spindle perpendicular to its base or table.

Vise - A mechanical device for holding a piece of metal rigid while it is being worked on.

Washer - A ring placed under a bolt or nut to distribute pressure.

Windlass - A device for hoisting weights that consists of a horizontal cylinder or drum turned by a lever or crank. A rope or cable winds around the cylinder or drum as the crank is turned and the weights are lifted to the desired position.

Workpiece - The piece of material that is in the process of being manufactured.

Worm gear - A type of gear that has teeth that resemble a thread. It is used to transmit motion between two shafts. The shafts are located at right angles to one another.

ODD-NUMBERED ANSWERS

SECTION 1 WHOLE NUMBERS

Unit 1 ADDITION OF WHOLE NUMBERS

Review Problems

a. 280 mm
b. 1,610 sq. in.
c. 15,906 ft.
d. $8,127
e. 60,490 km

Practical Problems

1. 6 302 kg
3. 2,611 bars
5. A = 200 mm
   B = 330 mm
   C = 580 mm
   D = 380 mm
7. a. 100 mm
   b. 65 mm
   c. 115 mm
9. a. Shaft 1: 18″
   Shaft 2: 35″
   Shaft 3: 43″
   Shaft 4: 33"
   Shaft 5: 32″
   b. 161″

Unit 2 SUBTRACTION OF WHOLE NUMBERS

Review Problems

a. 122 cm
b. 1,366 yd.
c. 8,292 oz.
d. $28,881
e. 2,466 ft.

Practical Problems

1. 187 m
3. 60 L
5. 60 mm
7. 60 mm
9. A = 6″
   B = 4″
   C = 2″
   D = 6″
11. 35″

Unit 3 MULTIPLICATION OF WHOLE NUMBERS

Review Problems

a. 1,116
b. 43,384
c. 894,864
d. $39,480
e. $1,614,800

Practical Problems

1. 74″
3. a. 42 m
   b. 153 m
   c. 24 m
   d. 319 m
5. a. 705 hr.
   b. 470 hr.
   c. 1,175 hr.
   d. 1,645 hr.
7. 39,648 lb.
9. 92,960 spindles
11. 48 times

Unit 4 DIVISION OF WHOLE NUMBERS

Review Problems

a. 6
b. $183\frac{18}{22}$
c. $16\frac{3}{17}$
d. 125
e. $11\frac{166}{789}$
f. $30\frac{9}{33}$

Practical Problems

1. 32 mm
3. 21,053 pieces/hr.
5. 372 parts
7. 40 parts
9. 16 pins
11. 15 reamers
13. 80 coil springs
15. a. $48
    b. $4

Section 1 CRITICAL THINKING PROBLEMS

1. Total weight = 6.3# x 5 = 31.5#
   $5.00 for first 10#
   21.5# x .35 = $7.53
   $5.00 + $7.53 = $12.53 to ship five parts

SECTION 2 COMMON FRACTIONS

Unit 5 ADDITION OF COMMON FRACTIONS

Review Problems

a. $\frac{1}{2}$
b. 1
c. $1\frac{1}{4}$
d. $1\frac{1}{6}$
e. $1\frac{5}{12}$

Practical Problems

1. $4\frac{7}{8}''$
3. $5\frac{1}{8}''$
5. $6\frac{9}{16}''$
7. $3\frac{1}{4}''$
9. a. $3\frac{31}{32}''$
   b. $4\frac{19}{32}''$
11. $3\frac{3}{16}''$
13. $7\frac{13}{32}''$
15. $8\frac{5}{64}''$
17. $3\frac{25}{64}''$

Unit 6 SUBTRACTION OF COMMON FRACTIONS

Review Problems

a. $\frac{1}{4}$
b. $\frac{3}{8}$
c. $\frac{1}{12}$
d. $\frac{1}{32}$
e. $\frac{21}{32}$

Practical Problems

1. $\frac{1}{2}''$
3. $\frac{19}{32}''$
5. $1\frac{11}{64}''$
7. $4\frac{5}{16}''$
9. $\frac{15}{32}''$
11. $1\frac{31}{32}''$
13. $1\frac{9}{16}''$
15. $8\frac{7}{32}''$

Unit 7 MULTIPLICATION OF COMMON FRACTIONS

Review Problems

a. $\frac{1}{8}$
b. $\frac{2}{9}$
c. $\frac{3}{16}$
d. 2
e. $15\frac{15}{16}$

Practical Problems

1. $29\frac{3}{8}''$
3. $73\frac{5}{16}''$
5. $6\frac{3}{4}''$
7. $197\frac{13}{16}''$
9. $26\frac{5}{8}''$
11. a. 59 teeth
    b. 66 teeth

Unit 8 DIVISION OF COMMON FRACTIONS

Review Problems

a. $1\frac{1}{2}$
b. $\frac{15}{16}$
c. $\frac{4}{9}$
d. $\frac{85}{96}$
e. $\frac{35}{51}$

Practical Problems

1. 4 reamers
3. a. 16 threads/in.
   b. $\frac{1}{16}''$
5. 136 revolutions
7. $4\frac{267}{364}$ lb.
9. $\frac{7}{16}''$
11. $\frac{7}{32}''$

Unit 9 COMBINED OPERATIONS WITH COMMON FRACTIONS

Practical Problems

1. 12 studs
3. 29 studs
5. $18\frac{3}{4}$ ft.
7. $24\frac{3}{16}''$
9. 53 punches
11. $6\frac{15}{64}''$
13. $62\frac{31}{32}''$
15. 20 ft. bar

Section 2 CRITICAL THINKING PROBLEMS

1. $44\frac{5}{32}''$

SECTION 3 DECIMAL FRACTIONS

Unit 10 ADDITION OF DECIMAL FRACTIONS

Review Problems

a. 4.4
b. 195.739
c. $24,911.19

Practical Problems

1. 129.75 mm
3. 6.875″
5. 8.402″
7. 458.23 mm
9. 70.359 mm
11. 88.874 mm
13. 60.325 mm

Unit 11 SUBTRACTION OF DECIMAL FRACTIONS

Review Problems

a. 53.3
b. 9585.277
c. $124.78

Practical Problems

1. 57.15 mm
3. 75.819 mm
5. 5.722″
7. 3.525″
9. 16.967 mm
11. 66.446 mm

Unit 12 MULTIPLICATION OF DECIMAL FRACTIONS

Practical Problems

1. 28.498,8 mm
3. 25.862 mm
5. 152.367,6 mm
7. 1.9″
9. 198.8″
11. 3.536 lb.

Unit 13 DIVISION OF DECIMAL FRACTIONS

Practical Problems

1. a. 8; 20
   b. 16; 30
   c. 20; 255
   d. 32; 116
   e. 4; 80
   f. 0.625; 65
   g. 0.8; 38
   h. 0.312,5; 18
3. 54.230 L
5. 0.1875″
7. a. 1.5915″
   b. 4.3768″
   c. 6.8436″
   d. 290.457,1 mm
   e. 220.429,1 mm
   f. 40.441,2 mm
9. a. 0.7647
   b. 0.7778
   c. 0.5833
   d. 0.8462
   e. 0.2727
   f. 0.8639
11. $^{20}/_{64}$″
13. $^{28}/_{32}$″

Unit 14 COMBINED OPERATIONS WITH DECIMAL FRACTIONS

Practical Problems

1. 7.625″
3. a. 4.5025″
   b. 2.255″
5. 17.14 mm
7. 5.747″
9. 1.55″
11. 0.2315″
13. 1.760″
15. 3.9283 lb.

Section 3 CRITICAL THINKING PROBLEMS

1. 3.5650 maximum
2. 3.5615 minimum

SECTION 4 DIRECT MEASURE

Unit 15 LENGTH MEASURING INSTRUMENTS

Practical Problems

1. a. $\frac{13}{50}''$
   b. $\frac{44}{50}''$
   c. $\frac{13}{100}''$
   d. $8\frac{1}{16}''$
   e. $8\frac{7}{32}''$
   f. $9\frac{3}{8}''$
   g. $8\frac{39}{64}''$
   h. $8\frac{51}{64}''$
   i. $9\frac{19}{64}''$
   j. $\frac{15}{16}''$
   k. $4\frac{3}{4}''$
   l. 1.6 cm or 16 mm
   m. 2.2 cm or 22 mm
   n. 3 cm or 30 mm
   o. 3.5 cm or 35 mm
3. a. 0.3233″
   b. 0.338″
   c. 12.55 mm
   d. 0.187″
   e. 7.522 mm
   f. 0.2345″
   g. 7.002 mm
   h. 5.00 mm
   i. 0.209″
   j. 9.25 mm

Unit 16 LENGTH MEASURE

Practical Problems

1. 36″
3. $17\frac{3}{4}''$
5. $11'\text{-}10\frac{51}{64}''$
7. $2'\text{-}7\frac{51}{64}''$
9. $6'\text{-}7\frac{61}{64}''$
11. 100 mm
13. 100 cm
15. 850 mm
17. 225 mm
19. 25 mm
21. 130 mm
23. 25 pieces

Unit 17 EQUIVALENT UNITS OF LENGTH MEASURE

Practical Problems

1. 127 mm
3. 95.25 mm
5. 0.776″
7. 1.722″
9. 0.854″
11. 5.075 dm
13. 7′-4″
15. 57.309 dm
17. 1,009.65 mm
19. 9 pieces
21. 1.587,5 mm
23. 16′-2$\frac{1}{2}$″
25. 1.685″

Unit 18 ANGULAR MEASUREMENT

Practical Problems

1. 90°
3. 3
5. 150°
7. 67° 13′
9. 67° 30′
11. 24°
13. 60°
15. 60°
17. 120°
19. a. 35°
    b. 110°
21. 30°

Section 4 CRITICAL THINKING PROBLEMS

1. Dime = $^{45}/_{64}$
   Nickel = $^{54}/_{64}$
   Quarter = $^{61}/_{64}$
3. a = 60°
   c = 6.33″

SECTION 5 COMPUTED MEASURE

Unit 19 SQUARE MEASURE

Practical Problems

1. 1,152 sq. in.
3. 6.5 sq. ft.
5. 0.165 dm^2
7. 1,348.5 sq. in.
9. 8.934 sq. ft.
11. 0.007,1 m^2
13. 79.354,7 cm^2

Unit 20 AREA OF SQUARES, RECTANGLES AND PARALLELOGRAMS

Practical Problems

1. 133 sq. ft.
3. $63.58
5. a. 696,772.8 mm^2
   b. 1,172,900.88 mm^2
7. 18.75 sq. ft.
9. 67 sq. in.

Unit 21 AREA OF TRIANGLES AND TRAPEZOIDS

Practical Problems

1. 2,074.5 sq. in.
3. 45.963 sq. in.
6. 1.953 sq. in.

Unit 22 AREA OF CIRCULAR FORMS

Practical Problems

1. 69,272.28 mm^2
3. 6,675.9 mm^2
5. 150.797 sq. in.

Unit 23 AREA OF CYLINDRICAL FORMS

Practical Problems

1. 15.904 dm^2
3. 2,709.63 cm^2
5. 56.5488 sq. in.

Unit 24 VOLUME OF RECTANGULAR SOLIDS

Practical Problems

1. 144 cubes
3. 1,728 cu. in.
5. 6.75 cu. in.
7. 241,300 mm^3
9. a. 1,350,000 mm^3
   b. 345,000 mm^3
   c. 1,875,000 mm^3
   d. 51,000 mm^3
   e. 589,934.304 mm^3
11. 0.2461 cu. in.

Unit 25 VOLUME OF CYLINDRICAL SOLIDS

Practical Problems

1. 68,722.5 mm^3
3. 13.3457 cu. in.
5. 12.25 cm
7. 16,022,160 mm^3
9. 0.0565 m^3
11. 248,873.625 mm^3
13. 10.9956 cu. in.

Unit 26 MASS (WEIGHT) MEASURE

Practical Problems

1. 141.855 lb.
3. 199.233 kg
5. 189 kg

Unit 27 VOLUME OF FLUIDS

Practical Problems

1. 0.792516 gal.
3. 53.86 gal.
5. 29.091 gal.

Section 5 CRITICAL THINKING PROBLEMS

1. Outside diameter area of pulley loop
   12 x 36 x 3.14 = 1356.5 sq. in.
   Inside area of pulley loop = 1356.5 sq. in.
   Outside edges of loop
   3.14 x 36 x .5 x 2 = 113.4 sq. in.
   Both sides of the center
   3.14 x 306.25 x 2 = 1923.3 sq. in.
   Area of center hole
   3.14 x 9 x 2 = 56.5 sq. in.
   Inside edge of center hole
   3.14 x 6 x .5 = 9.4 sq. in.
   Outside edge of center
   3.14 x 36 x .5 = 56.6 sq. in.

   1356.5 + 1356.5 + 113.4 + 1923.3
- 56.5 + 9.4 - 56.5 = 4646.1 sq. in.
= 32.3 sq. ft.

   32.3 ÷ 350 = .092 gallon of paint
required.

3. V = Area x Length
   = 28.27 x 18
   = 508.86 cu. in.

5. Hint: Subtract area of rod from area of piston.
   Area of rod = 3.1416 x .75 x .75
   = 3.1416 x .5625
   = 1.77 sq. in.
   Area of piston = 28.27 sq. in.
   28.27 - 1.77 = 26.5 sq. in.
   Force = 26.5 x 600 PSI = 15900# = 7.95 tons

SECTION 6 PERCENT AND GRAPHS

Unit 28 PERCENT

Practical Problems

1. a. 85%
   b. 3%
   c. 15%
3. a. $\frac{2}{5}$
   b. $\frac{2}{3}$
   c. $\frac{3}{4}$
5. 3 castings
7. 9 castings
9. a. 20%
   b. 80%
11. 22.92%
13. 84%
15. a. 22.22 hp
    b. 21.55 hp
17. a. 292.31 lb.
    b. 208.45 lb.
    c. 1,000 lb.

Unit 29 INTERPRETING GRAPHS

Practical Problems

1. T
3. T
5. T
7. a. 34°F
   b. 1.1°C
9. $\frac{1}{4}$″
11. $\frac{11}{16}$″
13. 16″

Section 6 CRITICAL THINKING PROBLEMS

1. Toluene = 10.3# x .23 = 2.369#/gallon
   Xylene = 10.3# x .11 = 1.133#/gallon
   EB = 10.3# x .019 = .196#/gallon

3. Toluene = 0 ton
   1.133# xylene x .35 = .397# less
   1.133 # - .397# = .736# xylene/gallon
   6989 x .736#/2000 = 2.57 tons xylene
   .196# EB x .05 = .0098# less
   .196# - .0098# = .1862# EB/gallon
   6989 x .1862#/2000 = .65 ton EB

SECTION 7 RATIO AND PROPORTION

Unit 30 RATIO

Practical Problems

1. a. 2:1
   b. 1:2
3. 1:1
5. 13:12
7. 3:1
9. 16:13
11. a. 1:25
    b. 25:1

Unit 31 DIRECT PROPORTION

Practical Problems

1. 25 people
3. 3.141 in.
5. 17.775 kg
7. a. 56 min.
   b. 25 min.
   c. 30 min.
   d. 18 min.
   e. 11 min.
9. a. $13\frac{1}{3}$; 15; 5
   b. 8; 15; all
   c. $6\frac{2}{3}$; 15; 10
   d. $5\frac{5}{7}$; 21; 15
   e. $3\frac{7}{11}$; 33; 21
   f. $2\frac{1}{2}$; 16; 8
   g. $2\frac{2}{19}$; 19; 2
   h. 2; 15; all
   i. $1\frac{19}{21}$; 21; 19
   j. $1\frac{1}{9}$; 18; 2
   k. $\frac{8}{9}$; 18; 16
   l. $\frac{40}{49}$; 49; 40
   m. $\frac{20}{27}$; 27; 20
   n. $\frac{2}{11}$; 33; 6
   o. $\frac{5}{47}$; 47; 5
11. 8 in.

Unit 32 INDIRECT PROPORTION

Practical Problems

1. a. 1,499 r/min
   b. 935 r/min
   c. 661 r/min
   d. 528 r/min
   e. 299 r/min

3. 275.7 mm

5. 32.1 in.

7. 326.7 rpm

9. 156 rpm

11. 8 teeth

13. a. 250 rpm counterclockwise
    b. 187.5 rpm clockwise
    c. 50 rpm counterclockwise

Section 7 CRITICAL THINKING PROBLEMS

1. $\frac{350}{X} = \frac{76}{118}$
   $76X = 350(118)$
   $X = 543.42$ ipm

3. $\frac{11.9}{16} = \frac{5.69}{X}$
   $16X = 67.71$
   $X = 4.23$ fpm

SECTION 8 SHOP FORMULAS

Unit 33 THREAD CALCULATIONS

Practical Problems

1.

	p	$f_{1,2}$	f_3	d_1	d_2
$\frac{5}{8}$–11	0.091 in.	0.011 in.	0.023 in.	0.056 in.	0.049 in.
M10 x 1.5	1.5 mm	0.188 mm	0.375 mm	0.920 mm	0.812 mm
$\frac{1}{2}$–20	0.050 in.	0.006 in.	0.013 in.	0.031 in.	0.027 in.
M6 x 1	1.0 mm	0.125 mm	0.250 mm	0.613 mm	0.541 mm
$\frac{5}{16}$–24	0.042 in.	0.005 in.	0.011 in.	0.026 in.	0.023 in.
M18 x 2.5	2.5 mm	0.313 mm	0.625 mm	1.534 mm	1.353 mm
M12 x 1.75	1.75 mm	0.219 mm	0.438 mm	1.074 mm	0.947 mm
1–12	0.083 in.	0.010 in.	0.021 in.	0.051 in.	0.045 in.

3. a. 0.0444"
   0.0692"
   0.0431"
   b. 0.0400"
   0.0620"
   0.0386"
   c. 0.0361"
   0.0563"
   0.035"
   d. 0.0227"
   0.0354"
   0.0220"
   e. 0.0546"
   0.0850"
   0.0529"
5. 0.138"
7. #5
9. 0.047"
11. 0.216"
13. 0.086"
15. a. 0.20"
    b. 0.1"
    c. 0.074"

Unit 34 GEAR COMPUTATIONS

Practical Problems

1. a. 6"
   b. 8 teeth/in.
   c. 0.125"
   d. 0.156"
   e. 0.031"
   f. 0.196"
   g. 0.393"
   h. 0.281"
   i. 0.250"
   j. 5.688"
3. a. 46 teeth
   b. 0.125"
   c. 0.156"
   d. 0.031"
   e. 0.196"
   f. 6"
   g. 5.438"
   h. 0.281"
   i. 0.25"
   j. 0.393"
5. a. 5.499,1 mm
   b. 1.750 mm
   c. 77 mm
   d. 2.749 mm

Unit 35 SPEED AND FEED CALCULATIONS FOR CYLINDRICAL TOOLS

Practical Problems

1. 23 rpm
3. 76 rpm, 31 rpm
5. 143.24 mm
7. a. 48 rpm
   b. 55 rpm
9. a. 46 rpm
   b. 6.04 min.
11. 0.32 min.

Unit 36 TAPER CALCULATIONS

Practical Problems

1. 1.697"
3. 1.0286"
5. 6.630"
7. 6.96"
9. a. 0.482"
   b. 0.0547"
11. 0.2917"

Section 8 CRITICAL THINKING PROBLEMS

1. $^{20}/_{1}$ = 20 tpi Possible Standard threads: $^{7}/_{16}$-20 or $^{1}/_{2}$-20
3. T= $^{6}/_{64}$(.040) = 2.34 minutes per cut
   2.34 x 10 passes = 23.4 minutes
5. (23.4+2.5) 100 = 2590$^{min.}/_{60}$ = 43.2 hrs.
   43.2 hrs. x $25 = $1,080.00 labor
   $440.32 x 100 = $44,032.00 for material
   $44,032.00 + $1,080.00 = $45,112 total

SECTION 9 POWERS, ROOTS, AND EQUATIONS

Unit 37 POWERS

Practical Problems

1. 125
3. 0.0049
5. 1.4025517307
7. 33.183,15 cm^2
9. 198.057,8 mm^2
11. a. 3^9
    b. N^6
    c. p^9
    d. 19^7
13. a. a^6
    b. 7^7
    c. B
    d. 5^7

Unit 38 SQUARE ROOTS

Review Problems

a. 2
b. 5.745
c. 8.832
d. 5.897
e. 11.180
f. 32

Practical Problems

1. a. 2
   b. 5
   c. 8
3. 1,111
5. $^{2}/_{3}$
7. $^{3}/_{5}$
9. 6.074,4 cm
11. 0.657,0 m

Unit 39 EXPRESSIONS AND EQUATIONS

Practical Problems

1. 12
3. 9
5. 7
7. 5
9. 29
11. $x = 6$
13. $x = 24$
15. $n = 17$
17. $a = 4.5$

Section 9 CRITICAL THINKING PROBLEMS

1. a. 9.3×10^{7}
   b. 1.86×10^{5}
   c. 1.0×10^{-3}
   d. 6.31×10^{-8}
   e. 3.77×10^{-5}
   f. 6×10^{0} (note: 10^{0} is defined as 1.)
   g. 2.92×10^{0}

SECTION 10 GEOMETRIC FORMS AND CONSTRUCTION

Unit 40 CIRCLES AND POLYGONS

Practical Problems

1. 141.42 mm
3. a. 18.12 mm
   b. 22.20 mm
   c. 22.50 mm
   d. 22.63 mm
5. a. 1.732"
   b. 1.000"
   c. 3.464"
   d. 2.000"
   e. 3.000"
7. a. 19.506 mm
   b. 35.389 mm
   c. 25.495 mm
9. a. 60°
   b. 109.94 mm
   c. 108.17 mm
11. 0.01745 RAD
13. 0.7854 RAD
15. 2.49583 RAD

Unit 41 PERIMETERS AND BEND ALLOWANCES

Practical Problems

1. 4,559.74 mm
3. 457.06 mm
5. a. P = 221.68 mm
   73.82 kg/mm^2
   BP = 25 megagrams
   Press 1
   b. P = 220 mm
   s = 30.23 kg/mm^2
   BP = 47 megagrams
   Press 3
   c. P = 350 mm
   s = 17.58 kg/mm^2
   BP = 31 megagrams
   Press 2
7. 671.91 mm

Unit 42 GEOMETRIC CONSTRUCTION

Practical Problems

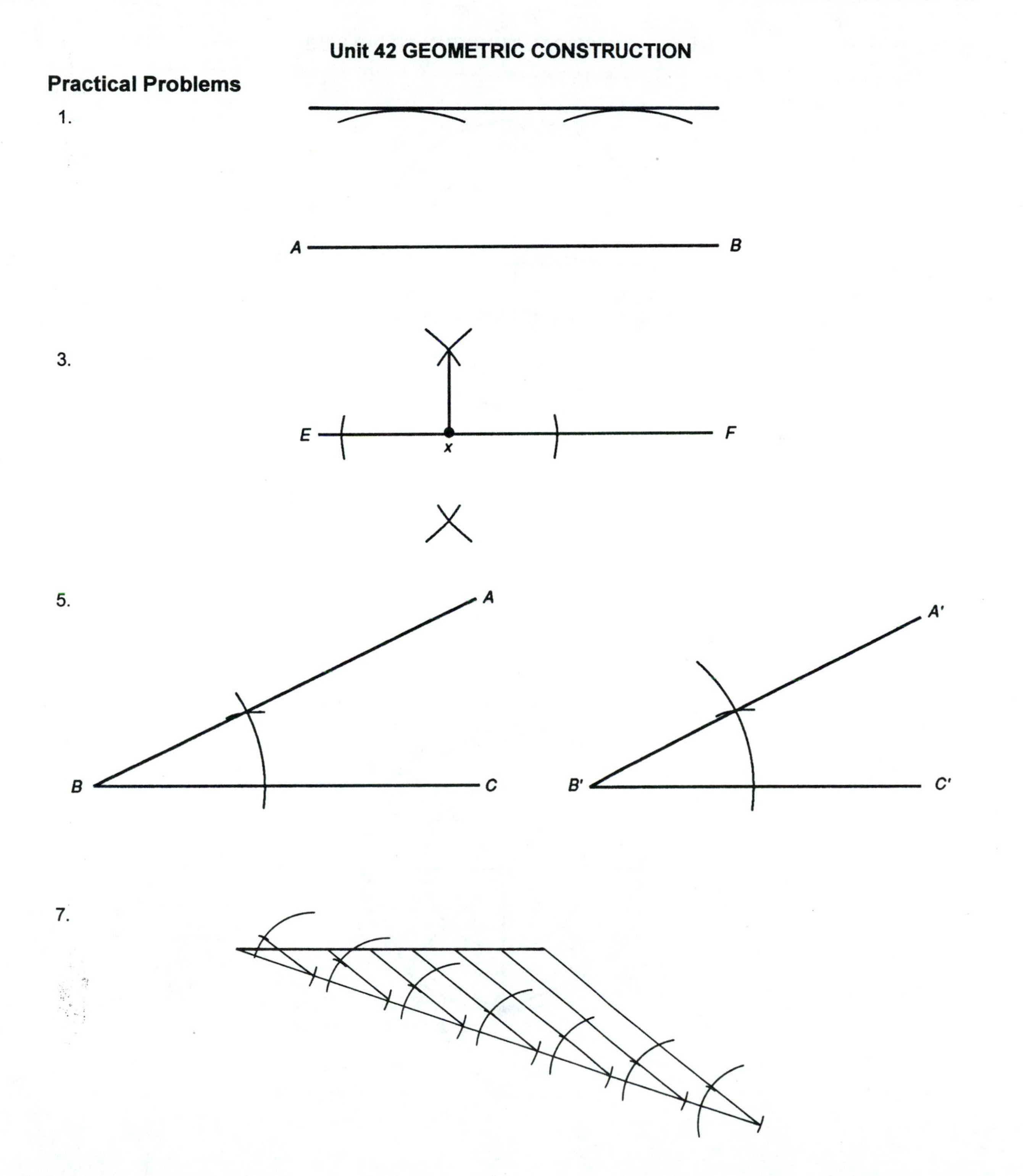

9. 

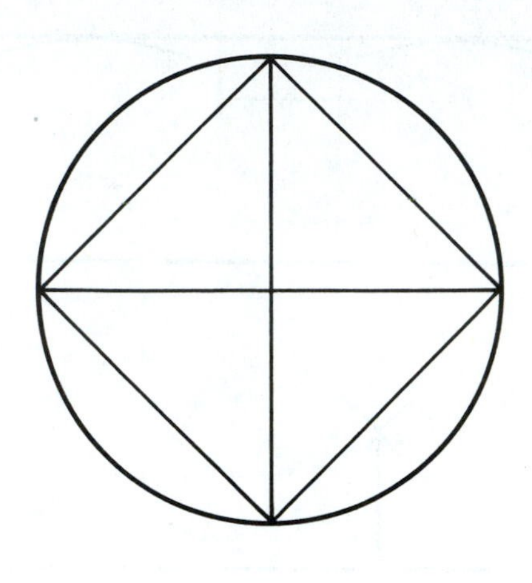

11.

13.

15.

17.

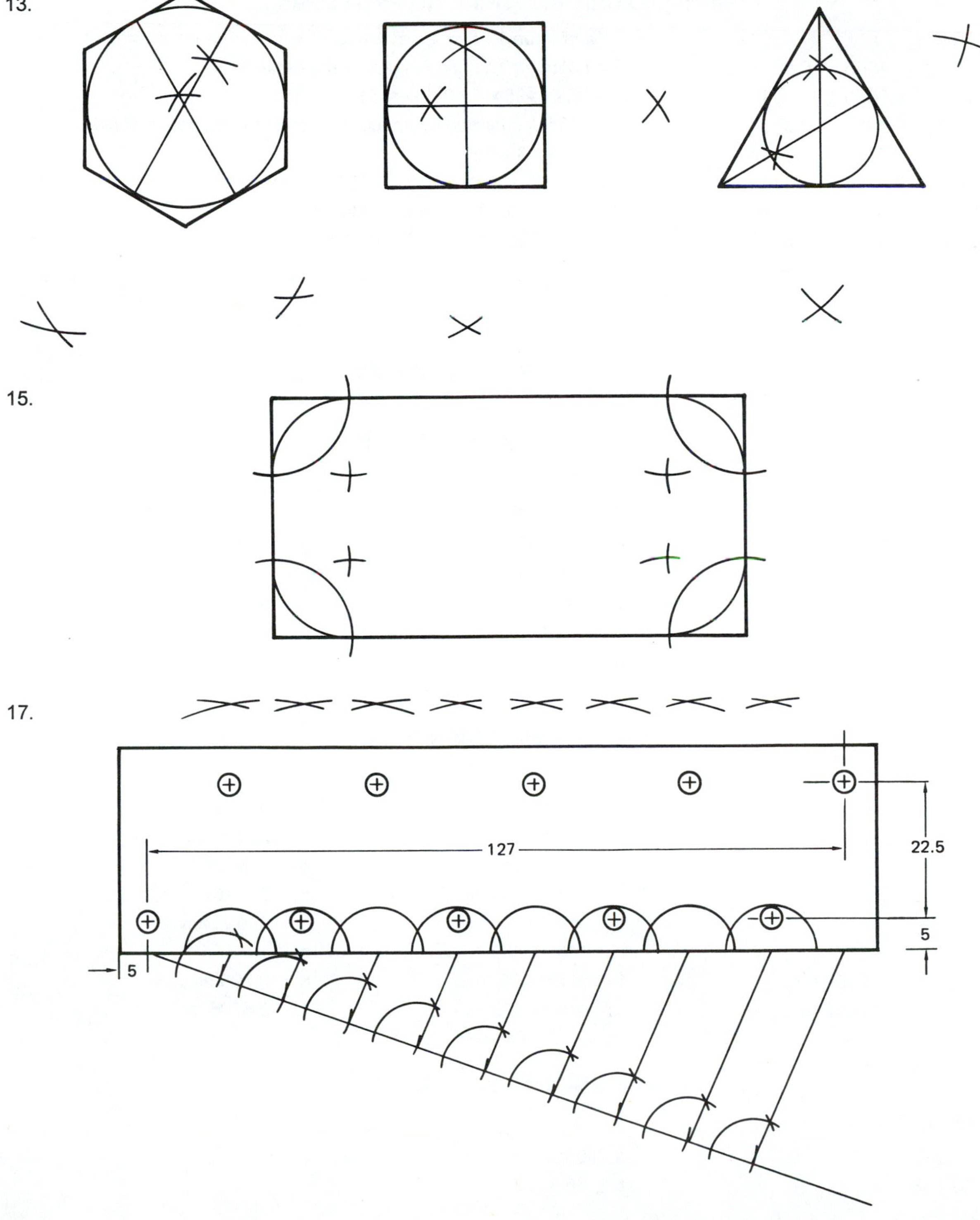

Section 10 CRITICAL THINKING PROBLEMS

1.

Hole #	x	y
1.	7.7546	5.1207
2.	8.9885	4.6096
3.	9.4996	3.3757
4.	8.9885	2.1417
5.	7.7546	1.6306
6.	6.5206	2.1417
7.	6.0095	3.3757
8.	6.5206	4.6096

3. Area of three 12-deg. arcs: $A = \pi(r^2) = 3.1416 \times 1 \times 1 = 3.1416$ in.^{2}
   Area of Inner Triangle formed by arc centers:
   $A = \frac{1}{2}BH = .5 \times 4 \times 3.464 = 6.928$ in.^{2}
   Area of sides formed by arc centers and tangent points of sides:
   $A = 1 \times 4 \times 3 = 12$ in.^{2}
   Area (total) = 3.1416 in.^{2} + 6.928 in.^{2} + 12 in^2 = 22.0696 in.^{2}
   Volume = 22.0696 in.^{2} x 0.1793 in. = 3.957 in.^{3}
   Weight = 3.957 in.^{3} x .283#/in.^{3} = 1.12#

SECTION 11 TRIGONOMETRY

Unit 43 NATURAL FUNCTIONS

Practical Problems

1. a. 0.89879
   b. 0.99768
   c. 0.75050
   d. 0.49278
   e. 0.60228
   f. 3.0992

3. a. 0.36197
   b. 0.36213
   c. 0.25504
   d. 0.65361
   e. 0.98413
   f. 0.99196
   g. 0.55156
   h. 0.87268
   i. 0.54275
   j. 0.43738

Unit 44 RIGHT TRIANGLES

Practical Problems

1. a. sine
   b. 44.44 mm
3. 101.71 mm
5. 1.81"
7. a. X = 4.177"
      Y = 4.480"
   b. X = 294.309 mm
      Y = 198.512 mm
   c. X = 0.655"
      Y = 13.985"
   d. X = 190.471 mm
      Y = 3.269 mm
   e. X = 3.785"
      Y = 11.650"
   f. X = 93.556 mm
      Y = 47.669 mm
9. 43.302 mm
11. a. 22.5 mm
    b. 3°34′35″
    c. 7°9′10″
    d. 360.7 mm
13. a. 10°
    b. 0.7053 in.
    c. 2.4106 in.

Unit 45 SINE BAR CALCULATIONS

Practical Problems

1. 1.3278 in.
3. 62.988 mm
5. a. 8°46′
   b. 38.103 mm
7. 22°54′

Unit 46 MEASURING ANGLES WITH DISCS

Practical Problems

1. a. 30°
   b. 0.3750″
   c. 0.6495″
   d. 1.0245″
   e. 5.1906″
3. 3.2255″
5. a. $\angle P = 60°$
   $D = 36.6$ mm
   b. $\angle P = 30°$
   $D = 11.4$ mm
   c. $\angle P = 66°$
   $D = 15.33$ mm
7. 151.74 mm
9. a. 4°10′36″
   b. 8.5757"

Unit 47 HELIC ANGLES

Practical Problems

1. a. 4°15′
   b. 3°41′
   c. 3°0′
   d. 2°55′
   e. 1°57′
   f. 2°40′
   g. 2°°8′
   h. 2°28′
3. a. 6°25′
   b. 4°40′
   c. 3°19′
   d. 3°19′
5. a. 8.908″
   b. 1,477.983 mm
   c. 336.857 mm
7. a. 568.490 mm
   b. 330.503 mm
   c. 4.825″
   d. 569.086 mm

Unit 48 ACUTE TRIANGLES

Practical Problems

1. a. 39.84 mm
   b. 23.00 mm
   c. 11.91 mm
   d. 44.43 mm
   e. 71.48 mm
   f. 19.15 mm
3. 0.0537″
5. 0.31428″

Unit 49 OBLIQUE TRIANGLES

Practical Problems

1.

	Sine	Cosine	Tangent
a.	+0.79863	-0.60181	-1.3270
b.	+0.51054	-0.85985	-0.59376
c.	-0.62501	-0.78061	+0.80067
d.	+0.98937	-0.14545	-6.8023
e.	-0.51931	+0.85459	-0.60767
f.	+0.61390	-0.78939	-.077768
g.	-0.87726	+0.48000	-1.8276
h.	+0.15228	-0.98834	-0.15407
i.	+0.27036	-0.96276	-0.28082

3. a. $a = 69.333$ mm
   $c = 84.679$ mm
   $\angle C = 62°2′$
   b. $a = 8.379$ mm
   $b = 14.936$ mm
   $\angle B = 32°0′11″$
   c. $b = 34.190$ mm
   $c = 41.376$ mm
   $\angle C = 64°28′38″$
   d. $a = 1{,}731.641$ mm
   $c = 3{,}243.941$ mm
   $\angle B = 55°27′$

Section 11 CRITICAL THINKING PROBLEMS

1. Tan 41 Deg. = .203/depth

   8692 = .203/depth

   depth = .203/.8682

   depth = .2335

3. Radius of bolt circle = 1.000"

   X = 2.000"

   Y = 1.500"